EXS 85

D-Amino Acids in Sequences of Secreted Peptides of Multicellular Organisms

Edited by P. Jollès

Birkhäuser Verlag
Basel · Boston · Berlin

Editor

Prof. Dr. P. Jollès
Laboratoire de Chimie
des Substances Naturelles
URA C.N.R.S. No. 401
Muséum National d'Histoire Naturelle
63, rue Buffon
F-75005 Paris
France

Library of Congress Cataloging-in-Publication Data
D-amino acids in sequences of secreted peptides of multicellular organisms / edited by P. Jollès.
p. cm. -- (EXS ; 85)
Includes bibliographical references and index.
ISBN 3-7643-5814-9 (hardcover : alk. paper). -- ISBN 0-8176-5814-9 (U.S. : hardcover : alk. paper)
1. Post-translational modification. 2. Enantiomers. 3. Amino acids. 4. Peptides. I. Jollès. Pierre, 1927– . II. Series.
QH450.6.D14 1998
572'.633--dc21

Deutsche Bibliothek Cataloging-in-Publication Data
D-amino acids in sequences of secreted peptides of multicellular organisms / ed. by P. Jollès. - Basel ; Boston ; Berlin : Birkhäuser, 1998
(EXS ; 85)
ISBN 3-7643-5814-9 (Basel …)
ISBN 0-8176-5814-9 (Boston)
EXS. - Basel ; Boston ; Berlin : Birkhäuser
Früher Schriftenreihe
Fortlaufende Beil. zu: Experientia

Printed on acid-free paper produced from chlorine-free pulp.TCF ∞
Printed in Germany
ISBN 3-7643-5814-9
ISBN 0-8176-5814-9

9 8 7 6 5 4 3 2 1

Contents

D-Amino acids hydrolysing enzymes

Homochirality and life

List of Contributors

M. Amiche, Laboratoire de Bioactivation des Peptides, Institut Jacques Monod, Université Paris 7, 2 Place Jussieu, F-75251 Paris Cedex 05, France

D. Barra, Istituto Pasteur-Fondazione Cenci Bolognetti, Dipartimento di Scienze Biochimiche "A. Rossi Fanelli" and CNR Centro di Biologia Molecolare, Università La Sapienza, Piazzale Aldo Moro 5, I-00185 Roma, Italy

W. A. Bonner, Department of Chemistry, Stanford University, Stanford, CA 94305, USA

F. Bossa, Dipartimento di Scienze Biochimiche "A. Rossi Fanelli" and CNR Centro di Biologia Molecolare, Università La Sapienza, Piazzale Aldo Moro 5, I-00185 Roma, Italy

A. Delfour, Laboratoire de Bioactivation des Peptides, Institut Jacques Monod, Université Paris 7, 2 Place Jussieu, F-75251 Paris Cedex 05, France

G. H. Fisher, Department of Chemistry, Barry University, 11300 Northeast Second Avenue, Miami, FL 33161-6695, USA

S. D. Heck, Pfizer Central Research, Eastern Point Road, Groton, CT 06340, USA

A. Huberman, Department of Biochemistry, Inst. Nac. de la Nutricion "Salvador Zubirán", Vasco de Quiroga 15, Tlalpan, Mexico City 14000, Mexico

D. Ingrosso, Institute of Biochemistry of Macromolecules, S.U.N. School of Medicine and Surgery, Via Constantinopoli 16, I-80138 Naples, Italy

Y. Kera, Department of Environmental Systems Engineering, Nagaoka University of Technology, Nagaoka, Niigata 940-21, Japan

G. Mignogna, Istituto Pasteur-Fondazione Cenci Bolognetti, Dipartimento di Scienze Biochimiche "A. Rossi Fanelli" and CNR Centro di Biologia Molecolare, Università La Sapienza, Piazzale Aldo Moro 5, I-00185 Roma, Italy

P. Nicolas, Laboratoire de Bioactivation des Peptides, Institut Jacques Monod, Université Paris 7, 2 Place Jussieu, F-75251 Paris Cedex 05, France

A. F. Perna, Institute of Biochemistry of Macromolecules, S.U.N. School of Medicine and Surgery, Via Constantinopoli 16, I-80138 Naples, Italy

A. Scaloni, Centro Internazionale di Servizi di Spettrometria di Massa-IABBAM, Consiglio Nazionale delle Ricerche, Via Pansini, I-80137 Napoli, Italy

M. Simmaco, Dipartimento di Scienze Biomediche, Università di Chieti G. D'Annunzio, Via dei Vestini, I-66100 Chieti, Italy

R.A. Volkmann, Pfizer Central Research, Eastern Point Road, Groton, CT 06340, USA

R. Yamada, Department of Environmental Systems Engineering, Nagaoka University of Technology, Nagaoka, Niigata 940-21, Japan

Y. Yasuda-Kamatani, Suntory Institute for Bioorganic Research, 1-1-1 Wakayamadai, Shimamoto-cho, Mishima-gun, Osaka 618-8503, Japan

Preface

Proteins are synthesized on polysomes as simple chains of L-amino acids. They can later be modified through numerous types of post-translational reactions. In fact, practically all their reactive groups can be modified *in vivo* by various enzymatic or chemical reactions. After the addition of carbohydrates, lipids, sulfate or phosphate groups, the properties of the primary products are changed in many regards, for example, with respect to solubility, stability, interactions of the proteins with receptor molecules, as well as catalytic and biological activities.

Recently, a new type of post-translational reaction has been described in sequences of peptides of multicellular organisms: the conversion of certain amino acids in peptide links from L- to D-configuration. This modification is quite subtle; as it cannot be detected by standard amino acid sequencing techniques and does not lead to a change in the molecular mass it can take place unnoticed.

It seems appropriate at this stage to summarize our current knowledge concerning the occurrence, the function and the biosynthesis of the D-amino acid containing peptides in *multicellular organisms*.

I sincerely thank the authors, all of them well-known experts in the field, for contributing incisive chapters to this book. The danger of redundancy is often present in works authored collectively. We have tried hard to minimize the problem as much as possible.

Pierre Jollès
Paris, February 1998

Characterization and analysis of D-amino acids

D-Amino Acids in Sequences of Secreted Peptides of Multicellular Organisms
ed. by P. Jollès

Characterization and analysis of D-amino acids

A. Scaloni[1], M. Simmaco[2] and F. Bossa[3]

[1] *Centro Internazionale di Servizi di Spettrometria di Massa-IABBAM, Consiglio Nazionale delle Ricerche, Via Pansini 5, I-80137 Napoli, Italy*
[2] *Dipartimento di Scienze Biomediche, Università di Chieti G. D'Annunzio, Via dei Vestini, I-66100 Chieti, Italy*
[3] *Dipartimento di Scienze Biochimiche "A. Rossi Fanelli" and CNR Centro di Biologia Molecolare, Università La Sapienza, Piazzale Aldo Moro 5, I-00185 Roma, Italy*

Summary. The need to screen a large number of natural extracts, with the aim of detecting D-amino acids or isolating and characterizing peptides containing them, has stimulated the development of novel and improved procedures for the analysis of amino acid enantiomeric mixtures, with special attention paid to automation. Different methods for the analysis of D-amino acids are described and discussed.

Introduction

D-Amino acids occur in living matter rarely as compared with their enantiomers, although their relative abundance is underestimated because of the inadequacy of available analytical procedures, both in terms of resolution and specificity. On the other hand, it is increasingly recognized that free and peptide-bound D-amino acids may play important roles in organisms. The majority of D-amino acids are found in antibiotic peptides synthesized by different microorganisms on special multienzymatic complexes [1]. In some cases, their presence in small quantities is considered to be the product of a racemization reaction related to molecular ageing [2]. As result of this phenomenon, racemization rates of amino acids have been used for dating ancient samples [3]. Free D-amino acids have been detected in several tissues of different origin, with enantiomeric mixtures being reported in humans [4, 5], rats [6], bivalve molluscs [7], marine invertebrates [8] as well as in microbially fermented foods and beverages as result of bacterial metabolism [9]. In addition, the presence of D-Asp and D-Ala has been detected alongside pathological conditions such as Alzheimer's disease [10]. Most interestingly, D-amino acids have also been observed in peptides and proteins of ribosomal origin [11]. Dermorphin, an opioid peptide from frog skin, was the first gene-encoded peptide found to contain a D-Ala at a specific position of the sequence, which was essential for biological activity [12]. This peptide is synthesized on the ribosome as a large precursor encoded by a messenger RNA (mRNA) where the triplet for L-Ala is present in the position occupied by the D-enantiomer in the mature product [13]. This L $\rightarrow$ D transformation is a result of novel posttranslational modification reactions. Subsequently, other D-amino acid-containing peptides were found in gene-encoded neuropeptides and antibiotics [14–17].

These factors have contributed to the need to develop automated systems to screen large numbers of natural extracts to detect D-amino acids or isolate and characterize active peptides containing these D-amino acids. These systems require development of novel procedures for the analysis of amino acid enantiomeric mixtures and to check stereochemical purity of synthetic peptides, with special attention paid to automation.

Enzymatic methods for D-amino acid analysis

Enzymatic methods, such as those based on L- or D-amino acid oxidases, have been used for 30 years [18], and an updated method was published very recently [19]. These procedures rely on the stereoselective loss of an amino acid after exposure to either oxidase as an indication of its configuration. Recently, another enzymatic procedure employing a bacterial D-amino acid transaminase was described [20], D-norleucine resulting from the transamination reaction, easily quantitated by conventional amino acid analysis. However, these procedures are all indirect and give reliable results only for amino acids that are good substrates for the enzymes. For these reasons, separation methods capable of resolving via simple analysis all L-amino acids from each other and from their corresponding enantiomers are preferable for direct determination of these molecules even when they are present at very low levels.

Separations techniques

Different separations techniques are available for the direct characterization of amino acid racemates. All methods exploit the general principles of either chiral stationary phases [21, 22], chiral mobile phases [23, 24] or precolumn derivatization with chiral reagents to obtain diastereomeric derivatives separable on conventional chromatographic supports [25, 26]. A number of different liquid chromatography (LC), gas chromatography (GC), supercritical fluid chromatography (SFC) and capillary electrophoresis (CE) applications to give chiral recognition have been reported in the literature [27, 28].

Chiral stationary phases

In the case of chiral stationary phases (CSPs), the selective separation of a racemic mixture is based on the preferential interaction of one of the two enantiomers with an optically active substance immobilized to a inert support. Nuclear magnetic resonance (NMR), molecular dynamics, liquid-solid chromatography and interface solvation studies have allowed the formulation of the theory of enantioselective complexation as a basis of the chiral recognition phenomena [29]. Different CSPs have been developed to maximize the difference of formation energy between the diastereomeric adsorbates that have to be separated (Fig. 1). Chiral polymers such as cellulose, carbamates or polyacrylamide have been used successfully to separate many organic enantiomers, but these have been totally unsatisfactory with amino acids [30, 31]. However, different commercially available proteins irreversibly bound to silica or agarose have been used in LC to resolve a variety of chiral amino acid derivatives. α-Chymotrypsin and albumin allowed the separation of free and derivatized amino acids [32, 33], but these kinds of supports were not further developed because of limited chromatographic resolution and protein degradation over time at the temperature and pH values used.

Optimization of hydrogen-bonding, π donor-acceptor, dipole-stacking and steric interactions between the CSP and analyte has made possible preparation of a number of highly selective CSPs suitable for a broad range of applications and showing high chromatographic efficiencies. These stationary phases, termed donor-acceptor (DA) CSPs, are designed on the basis of chiral recognition-modelling studies of the analyte and the chiral precursor that have to be immobilized on inert support. Thus, different chiral polysiloxanes such as Chirasil-Val found a very impressive GC application in the separation of *N*(*O*,*S*)-trifluoroacetyl-amino acid *n*-propyl esters [34, 35]. Eighteen racemic amino acids, previously derivatized with trifluoroacetic anhydride and esterified with *n*-propanol, were completely resolved within a 40-min analysis. After the original suggestion of a two-point hydrogen-bonding chiral recognition model [36], Pirkle proposed a mechanism based on a stacking π-dipole multipoint interaction between the substituents and the stereogenic center [37]. In fact, as illustrated in Figure 2 (panel A), two-point hydrogen bonding alone is incapable of giving rise to chiral recognition; a face-to-face interaction promoted by stacking π-dipoles possibly explains the observed chiral recognition, since it would juxtapose in proximity the functionalities of both chiral centers and could lead to the required stereochemically dependent interaction. In almost every case, successful DA-CSPs contain some aromatic or extended π-functionality, because π-π interactions are often a major controlling factor in enantiomer separation of racemic mixtures on these stationary phases. Different DA-CSPs have been successfully used for the LC analysis of amino acid enantiomers. Racemic mixtures are usually reacted with nonchiral precolumn derivatizing reagents to introduce a fluorogenic or ultraviolet (UV) absorbing moiety in the molecule, and then separated by DA-CSPs [38–42]. Figure 2 (panel B) reports the chemical structure of the chiral moieties of some Pirkle-type CSPs used for the separation of derivatized amino acids.

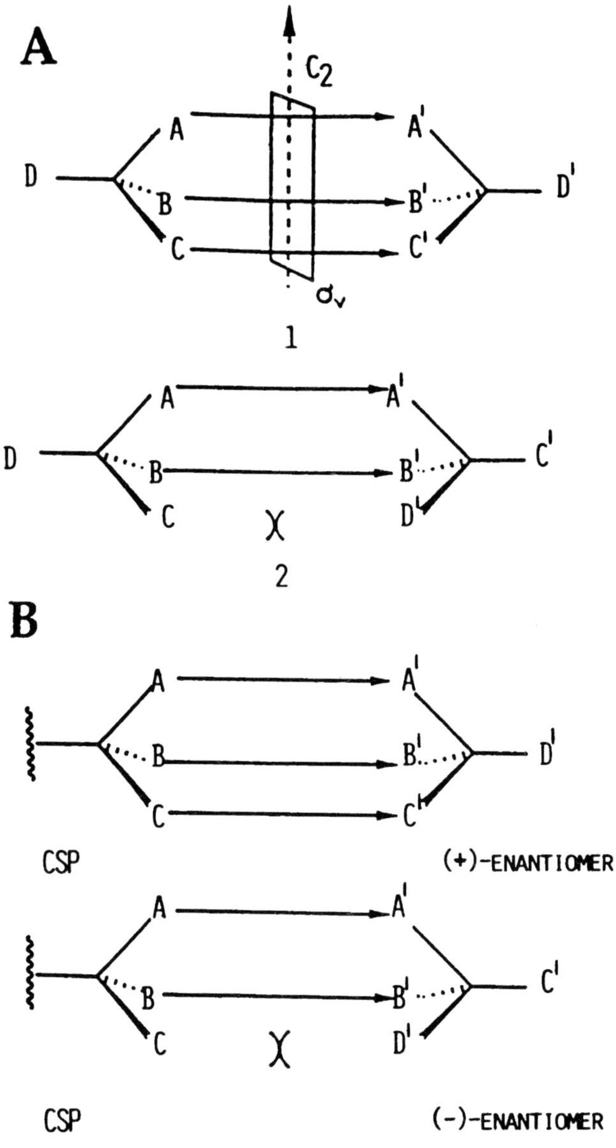

Figure 1. Enantiomeric recognition. (A) Diastereomeric complexes 1 and 2 differ in symmetry. Complex 1 has a plane of symmetry and a C2 axis of pseudosymmetry. These elements are absent in complex 2. If one of the species involved in complex formation is fixed to a stationary support, a CSP is generated as shown in (B). The (+) enantiomer of the analyte is capable of three simultaneous interactions with the CSP (A-A′, B-B′ and C-C′), whereas the (–) enantiomer is capable of only two simultaneous interactions. If all three interactions are free-energy-lowering, the (–) enantiomer will be less retained by the CSP. (From ref. 29).

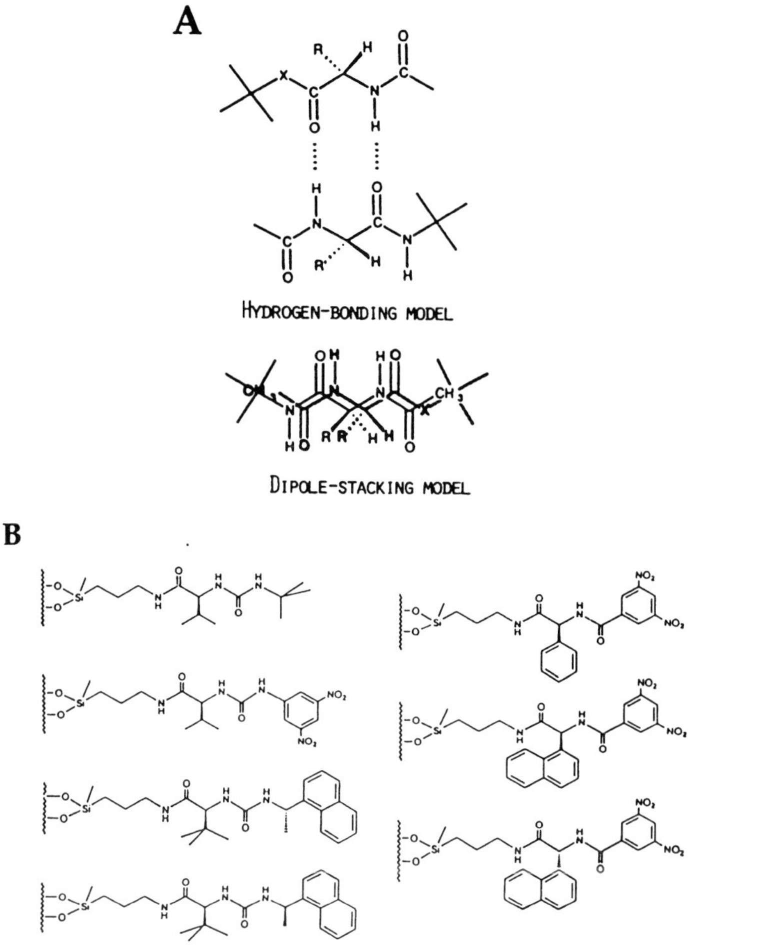

Figure 2. Dipole stacking multipoint interaction. (A) Hydrogen bonding vs. dipole stacking mechanism for chiral recognition of *N*-acyl-α-amino amides on *N*-formyl-valine derived CSP. (B) Chemical structures of Pirkle-type stationary phases used for the enantiomeric separation of derivatized amino acids. (From refs. 29 and 38).

These stationary phases bear (*S*)-Val amide, (*S*)-*tert*-Leu amide, (*S*)-phenyl-Gly amide, or (*S*)-naphthyl-Gly amide functionalities immobilized on silica-based stationary phases.

Chiral mobile phases

In the case of chiral mobile phases (CMPs), separation of the amino acid enantiomers on conventional LC is achieved by including a chiral additive in the eluants [43]. The stereoselectivity is ascribed to differences in stability of the diastereomeric amino acid-additive complexes formed in solution. Different techniques have been developed depending on the chiral selectors used. In ligand-exchange LC, the complexation of amino acids is obtained by the addition of a bidentate analyte enantiomer and metal ions (usually Cu^{2+} or Ni^{2+}). The separation depends on the differential partitioning of the diastereomeric complexes between the stationary phase and the mobile phase. The pioneering studies of Hare and Gil-Av described the separation of underivatized amino acid enantiomers on an ion-exchange column by the addition of L-proline and Cu^{2+} salts [23]. A simple change in the chirality of the eluant reversed the order of enantiomer elution. The same solvent additives allowed the separation of dansylated amino acids by conventional reversed-phase high-pressure liquid chromatography (HPLC) [44]. Figure 3 (panel A) describes the ternary complexes proposed in this case. The effect of other chiral additives such as L-amino acid derivatives, *N*-(1′-hydroxymethyl)-propyl-α-aminobenzylphosphonic acid derivatives, L-pipecolic acid, diamino-diamido type ligands and L-azetidine-2-carboxylic acid on the separation of racemic mixtures was investigated [45–48], but a complete separation of L-amino acids from each other and from their corresponding enantiomers was never obtained in a single LC run.

Other chiral selectors, such as chiral crown ethers, can be added to the mobile phase in order to obtain enantioselective separation of amino acids. These molecules present a cavity with a high degree of chiral recognition that can selectively host the analytes to be resolved. In addition to chiral crown ethers initially used, naturally occurring cyclodextrins (CDs) have found attractive applications in amino acid analysis [49]. CDs can be represented as a truncated cone with the secondary hydroxyl groups of the glucose units pointing to one side of the opening and the primary hydroxyl groups to the opposite side (Fig. 3, panel B). The interior of the cavity consists of –CH– groups providing a relatively hydrophobic character with respect to the hydrophilic cavity entrances. Thus, molecules of a certain size and stereochemistry can be included in the CD cavity by hydrophobic interactions. In fact, glucose units form a chiral cavity that functions as a chiral selector, producing different diastereomeric inclusion complexes with two enantiomeric analytes, which can be easily separated. Different LC and CE applications have been successfully developed for the analysis of amino acid enantiomers by CDs. Generally, racemic mixtures are reacted with nonchiral precolumn derivatizing reagents to introduce a fluorogenic or UV-absorbing moiety in the molecule, and then separated by CD-containing media [50, 51].

Ligand-exchange and cyclodextrin-bonded stationary phases were also developed: although the mechanism of chiral discrimination is the same as that described in this section, they can be classified as CSPs [52, 53].

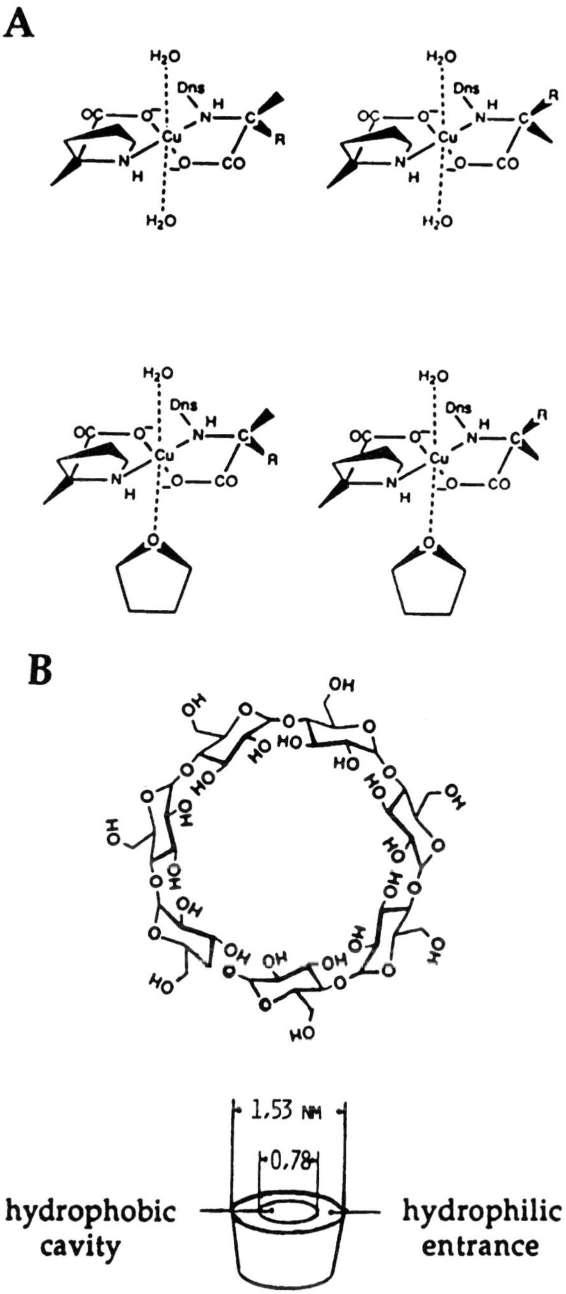

Figure 3. Proposed structures of the ternary complexes of L-proline and D,L-isomers of dansylated amino acids with copper(II) in coordination with water and water-tetrahydrofuran (A). β-Cyclodextrin (B): chemical structure and geometrical arrangement of the glucose units. (Reprinted from ref. 48, with kind permission from Elsevier Science – NL, Sara Burgerhartstraat 25, 1055 KV Amsterdam, The Netherlands, and ref. 90).

Precolumn derivatization with chiral reagents

The introduction of an additional chiral center on the analyte by precolumn derivatization with a chiral reagent results in the formation of different diastereomers with distinctive chemicophysical properties, thus separable on conventional reversed-phase supports. In addition, the derivatizing reagent introduces a fluorescent or UV-absorbing moiety essential for detection. For effective diastereoisomer formation, the reagent should be of high optical purity and racemization of the substrate or the reagent should not occur during the reaction. Mild reaction conditions in aqueous media are therefore generally employed. The amino group is the preferred site for derivatization, since modification of the carboxyl group can increase the risk of racemization.

The first methods described were based on the reaction of the amino group with chiral reagents such as *d*-10-camphorsulphonyl chloride [54], α-methoxy-α-methyl-naphthalenacetic acid [55], (–)1,7-dimethyl-7-norbornyl isothiocyanate or (+)neomenthyl isothiocyanate [56], followed by esterification of the carboxyl group with a chromophore, but poor results were obtained for the derivatives analyzed by reversed-phase chromatography. Acetylated sugar molecules were then reported for the resolution of amino acids. The reagent 2,3,4,6-tetra-*O*-acetyl-β-D-glucopyranosyl isothiocyanate was first used for the separation of amino acid esters [57] and later of free amino acids [58]. Excellent reversed-phase chromatography resolution was obtained for free amino acid derivatives, considerably better than for the esters. The related reagent 2,3,4-tri-*O*-acetyl-α-arabinopyranosyl isothiocyanate gave similar results, but the elution order was reversed, the D-enantiomer being eluted before the L one. The first separation of protein amino acid enantiomers in a single HPLC run was accomplished through derivatization with 2,3,4,6-tetra-*O*-acetyl-β-D-glucopyranosyl isothiocyanate [59], although peak resolution was not excellent.

Davis and Mohammed employed chiral fluorescent reagents, such as 9-fluorenylmethoxycarbonyl-L-amino acid isothiocyanates, for separation of free amino acids [60]. Their chiral derivatives were synthesized with the aim of studying the extent of racemization during the Edman sequencing procedure. Then, the reversed-phase HPLC (RP-HPLC) separation of fluorescent (+)1-(1-naphthyl)ethyl carbamoyl amino acid derivatives was described [61]. With this reagent, it was possible to detect 1 pmol of D-amino acid in the presence of 1000-fold excess of the corresponding L-isomer, although overlapping peaks in the chromatographic run were observed.

Several authors have described the use of *o*-phthaldialdehyde (OPA) as a fluorescent label: Aswad employed *N*-acetyl cysteine instead of the commonly used mercaptoethanol for labelling aspartic acid [62]; Buck and Krummen separated the enantiomers of 10 amino acids in 40 min by derivatizing with OPA and tBoc-cysteine [63]; Nimura and Kinoshita described the separation of 13 amino acids in about 70 min using OPA and

N-acetyl cysteine [64]. Other thiol reagents were then proposed in combination with OPA, such as the sodium salt of 1-thio-*β*-D-glucose, 2,3,4,6-tetra-*O*-acetyl-1-thio-*β*-D-glucopyranoside or (3*s*)-neomenthyl-thiol, but the chromatographic separation of the derivatives was not satisfactory [65]. In 1994, the use of *N*-isobutyryl-L-cysteine was proposed as a reagent for the simultaneous analysis of a 41-component standard mixture containing the 17 protein amino acids in racemic form, plus glycine and other non-protein amino acids [66] (Fig. 4). This fluorescent reagent allowed the detection of 1–2 pmol of each amino acid. The derivatives were completely separated in 75 min on a standard octadecylsilyl stationary phase.

The development of (+)-1-(9-fluorenyl)ethyl-chloroformate (FLEC) [67] was a direct consequence of the application of 9-fluorenylmethyl-

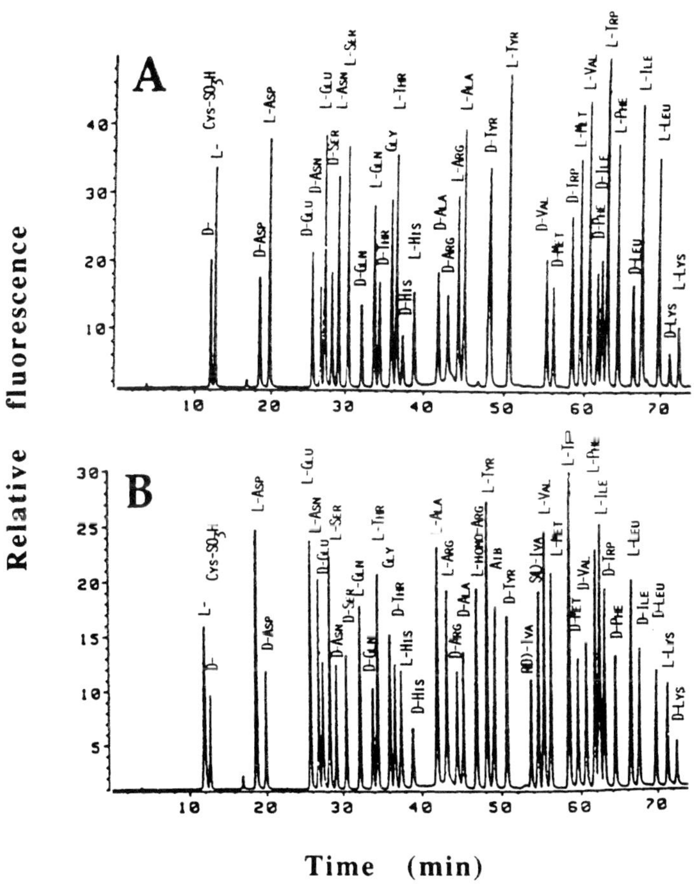

Figure 4. Elution profiles of amino acid standards derivatized with (A) OPA-isobutyryl-D-Cys and (B) OPA-isobutyryl-L-Cys. Amount analysed: 50 pmol of each L- and D-amino acid plus glycine. (Reprinted from ref. 66, with kind permission from Elsevier Science – NL, Sara Burgerhartstraat 25, 1055 KV Amsterdam, The Netherlands).

chloroformate to nonchiral amino acid analysis [68]. This optically active reagent yielded highly fluorescent amino acid diastereoisomers within 4 min incubation at room temperature without racemization. The derivatives displayed favorable chromatographic properties that allowed the reversed-phase separation of the D and L forms of 17 primary amino acids in a single run with only one case of coelution. The advantages with respect to the OPA-based reagents were the higher stability of the derivatives and the possibility of resolving secondary amino acids in a separate run. FLEC was also successfully used in micellar electrokinetic chromatographic applications [69]. The structures of the chiral reagents described above as well as the Marfey reagent [70] (*vide infra*) are shown in Figure 5.

Amino acids derivatized with Marfey reagent, 1-fluoro-2,4-dinitrophenyl-5-L-alanine amide (FDAA) can be separated into their D and L forms with elution volumes often differing by several minutes, using conventional RP-HPLC conditions. In the original version of the method [70], FDAA was used to derivatize five amino acid enantiomeric mixtures: aspartic and glutamic acids, alanine, methionine and phenylalanine. This compound efficiently reacts with the amino group of amino acids giving stable, yellow derivatives that absorb intensively at 340 nm. Sodium bicarbonate was used to provide the basic conditions necessary for the coupling reaction. More recent papers reported (i) the separation of FDAA diastereomers by a combined two-dimensional thin-layer chromatography (2D-TLC)/RP-HPLC procedure [6], (ii) the separation of racemic mixtures by the use of different buffers depending on the amino acid nature [71] or (iii) an RP-HPLC analysis limited to the L-amino acids present in protein hydrolysates [72]. In some cases, the conditions described gave overlapping peaks or required HPLC analyses of samples recovered from TLC.

We recently described a fully automated system for the analysis of amino acid enantiomeric mixtures using the commercially available Marfey reagent (L-FDAA) and its enantiomer 1-fluoro-2,4-dinitrophenyl-5-D-alanine amide [73]. The system automatically derivatizes and injects the sample onto the liquid chromatograph, allowing the analysis of enantiomeric primary and secondary amino acid mixtures continuously 24 h a day completely unattended. The HPLC profile of 15 pairs of enantiomeric amino acids plus glycine after precolumn derivatization with L-FDAA is shown in Figure 6. Each D-amino acid derivative is well separted from the corresponding L stereoisomer; only the coelution of L-Leu with D-Val is observed. In another study, D diastereomers were reported to be eluted from the column after the L ones as a result of a hypothesized intramolecular hydrogen bonding [74]. Similar results were obtained in our case, except for the histidine and arginine derivatives. If the derivatization is performed with 1-fluoro-2,4-dinitrophenyl-5-D-alanine amide, the elution order of the amino acid diastereomers is reversed. The use of the D form of the reagent allows an unequivocal determination also in the case of coelution between L-Leu and D-Val mentioned above. The linearity of the response for all the

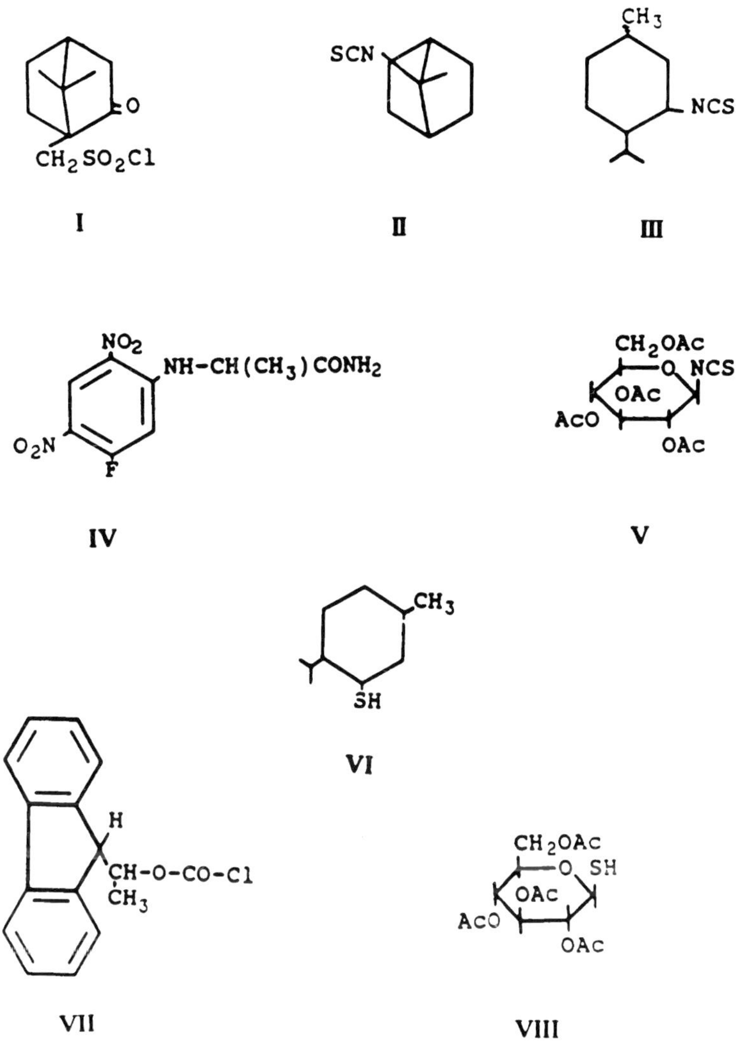

Figure 5. Precolumn derivatization reagents used for the resolution of enantiomeric mixtures of amino acids. (I), *d*-10-camphorsulphonyl chloride; (II), (−)1,7-dimethyl-7-norborynl isothiocyanate; (III), (+)neomenthyl isothiocyanate; (IV), 1-fluoro-2,4-dinitrophenyl-5-L-alanine amide (FDAA); (V), 2,3,4,6-tetra-*O*-acetyl-*β*-D-glucopyranosyl isothiocyanate; (VI), (3s)-neomenthyl-thiol; (VII), *o*-phthaldialdehyde; (VIII), (+)1-(9-fluorenyl)ethyl-chloroformate (FLEC); (IX), 1-thio-*β*-D-glucose.

amino acids analyzed was verified across the range 50–1000 pmol. The use of L-norvaline (eluting 0.8 min later than L-Val) as an internal standard allowed us to calculate the recovery for each sample. Thus, with both reagents each amino acid derivative can be detected and quantitated in the presence of a large excess of the corresponding diastereomer. Traditionally, precolumn derivatizations are performed offline from the liquid chro-

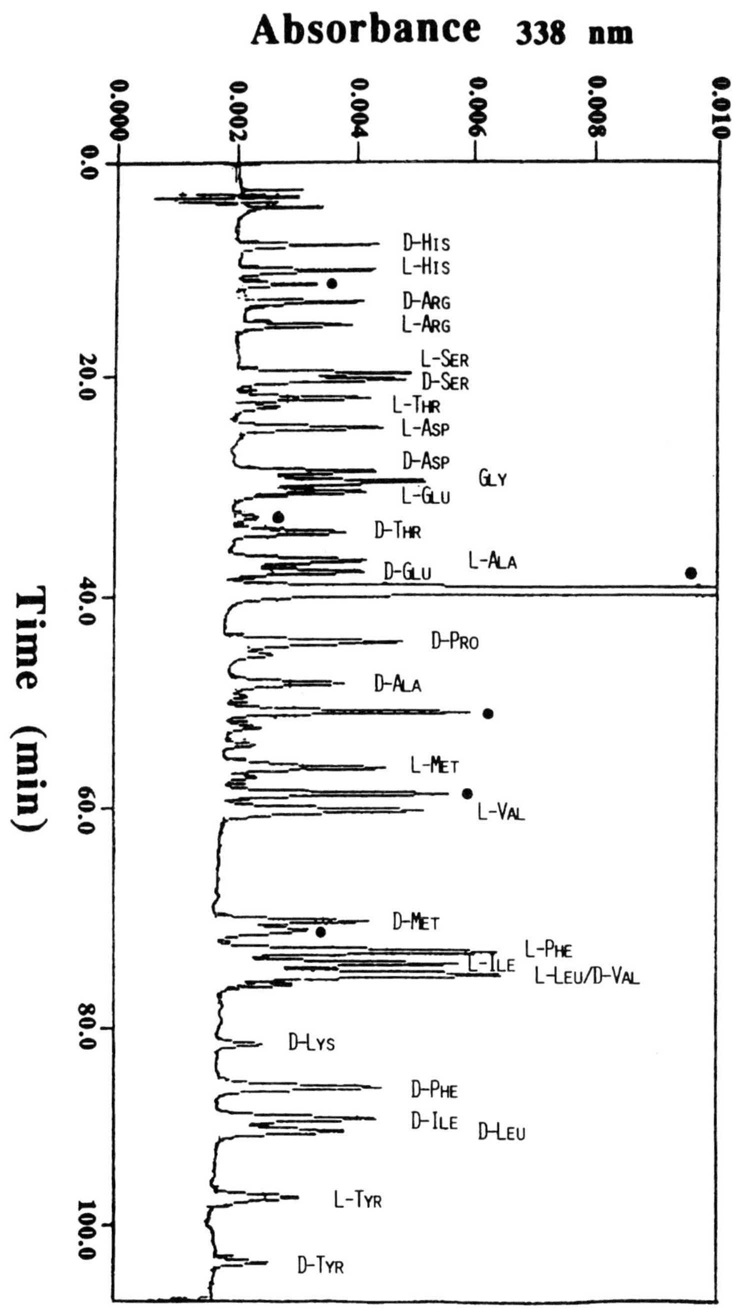

Figure 6. Reversed-phase HPLC elution profile of a standard mixture of D and L-amino acids plus glycine (150 pmol each) derivatized with L-FDAA. The peaks denoted with • are reagent-related. (From ref. 73).

matograph. They are therefore time-consuming, and errors inevitably occur when using manual procedures. Moreover, decomposition of sample derivatives takes place when analysis is delayed. The system described was based on the complete automation of both derivatization and injection. During the analysis of a sample, the next one is derivatized. Different thermostating of the reagents (5°C) and samples (50°C) allows the storage of the reagents without degradation. The substitution of triethylamine for sodium bicarbonate, originally used in the derivatization mixture, allows the drying of the sample directly in the reaction rack, avoiding the formation of nonvolatile salts during the acid-neutralization step. An *ad hoc* computer program to control the derivatization of amino acid racemates and sample injection was designed. More recently, different 1-fluoro-2,4-dinitrophenyl-5-L-amino acid amide analogs have been synthesized and checked for their capability to derivatize and chromatographically resolve enantiomeric mixtures. Among them, 1-fluoro-2,4-dinitrophenyl-5-L-valine amide seems to give the largest differences in retention times of each diastereoisomeric pair [74].

Most of the methods described prescribe the inclusion of nonvolatile buffers of low ionic strength and a strictly fixed pH value of the elution solvents. Thus, some problems may occur since slight modifications in these parameters produce significant shifts in retention times. Furthermore, adverse effect of salts on maintenance of hydraulic components of the instrumentation must be taken in consideration, particularly when an automated routine for the screening of a large number of samples is necessary. To overcome these problems, an alternative RP-HPLC procedure has been developed. The separation is carried out using a Spherisorb S5 ODS2 column, thermostatted at 50°C and eluted with a linear gradient from 25 to 50% of 70% acetonitrile/2-propanol (4:1, v/v) in 0.1% trifluoroacetic acid (TFA) at a flow rate of 0.4 ml/min. Accordingly, the buffer for the automated derivatization has been modified. A mixture of triethylamine/acetonitrile (1:1, v/v) increases the derivatization yield of basic amino acids and tyrosine as FDAA-di-derivatives. Furthermore, a mixture of acetonitrile/water/ethylacetate (1:1:0.01, v/v/v) in TFA 0.1% was found to better solubilize the FDAA derivatives before loading into the column. A typical separation of D and L hydrophobic amino acids is shown in Figure 7. This approach proved to be extremely helpful in the analysis of molecules belonging to the class of lipodepsipeptides [75], rich in D and L hydrophobic amino acid residues.

Finally, capillary electrophoresis and micellar electrokinetic separation of derivatives obtained with the Marfey reagent has been described [76]. However, this procedure did not separate all L-amino acids from each other and from their corresponding enantiomers in a single run.

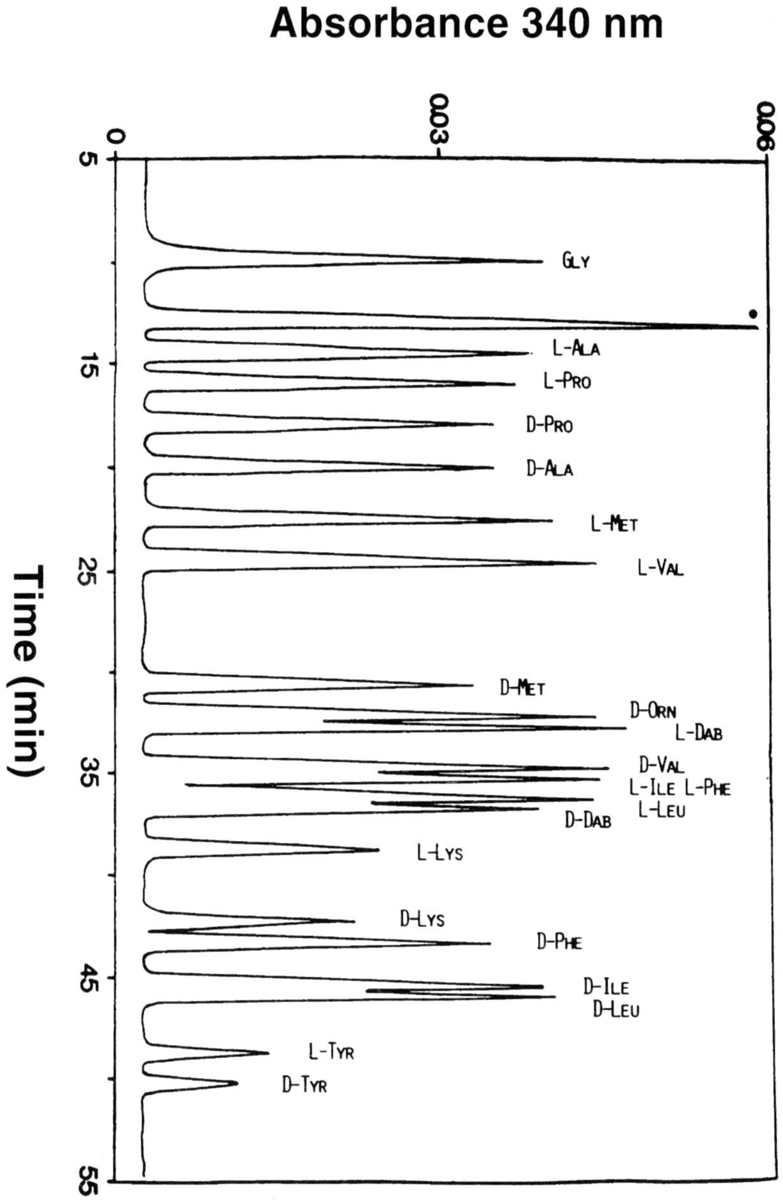

Figure 7. Reversed-phase HPLC elution profile of a mixture of D and L hydrophobic amino acids plus glycine (150 pmol each) derivatized with L-FDAA. The peak denoted with • is reagent-related.

Racemization during acid hydrolysis

Low levels of racemization are usually observed in the course of compositional analysis of peptides, depending upon the conditions used for acid hydrolysis [77]. In order to determine the chiral purity of the amino acids in a peptide, this phenomenon has to be corrected or taken into account. However, accurate correction factors are difficult to obtain, because racemization rates depend on the localization of an amino acid in the peptide and differ among individual free amino acids [78]. When we analyzed the compositional stereochemistry of natural or synthetic peptides, we always observed racemization values compatible with those reported in the literature. A recent study based on sample hydrolysis in deuterated acid, precolumn derivatization with a chiral reagent and analysis by HPLC-ES/MS used Marfey's reagent for the determination of the chiral purity of a synthetic peptide [79]. Each amino acid that racemizes during peptide hydrolysis is labelled with one deuterium atom at the α-carbon; thus, the mass increase of the racemized amino acids compared with the nonracemized ones is easily quantitated by mass spectrometry.

Determination of the chirality in the course of sequence analysis

When the same amino acid occurs repeatedly in the peptide sequence, it is desirable to determine the chirality of each residue as a function of position. Unfortunately, this cannot be done by determining the chirality of the phenylthiohydantoins obtained in the course of Edman degradation, since these have a marked tendency to racemize [60, 80]. In fact, the use of chiral isothiocyanates showed that the cleavage of thiocarbamoyl amino acids and their cyclization to the dihydrothiazolinones does not occur without racemization, because an equilibrium is established with a quasi-aromatic, nonchiral tautomeric form. Recently, when the new fluorogenic Edman reagent, 7-*N*,*N*-dimethylaminosulphonyl-4-(2,1,3-benzoxadiazolyl) isothiocyanate (DBD-NCS) was used for the analysis of [D-Ala^2, D-Leu^5]enkephalin [81], the epimerization of the thiazolinone derivative observed at each cycle of degradation was always above 20%. Although the use of chiral isocyanates largely suppressed the racemization, the yields of diastereomeric hydantoins were unsatisfactory, and their separation by HPLC incomplete. Bolte et al. [82] demonstrated that alkylcarbamoyl amino acid esters can be sequentially removed from the N-terminus of peptides with retention of configuration. The derivatives can subsequently be identified by gas chromatography on chiral columns. However, this method is not suitable for peptides containing lysine, histidine, arginine or tryptophan and requires relatively large amounts of starting material.

We previously described the synthesis and the use of a chiral compound, 1-fluoro-2,4-dinitrophenyl-5-L-alanine (FDNPA), for the determination of the chirality of amino acid residues in the course of sequence analysis of peptides [83]. Like the Marfey reagent, this compound efficiently reacts with the amino group of amino acids and peptides giving stable, yellow derivatives that absorb intensively at 340 nm ($\varepsilon = 4 \times 10^4$ M^{-1} cm^{-1}). The reversed-phase HPLC profile of 14 pairs of enantiomeric amino acids plus glycine and L-norvaline as internal standard, after precolumn derivatization with FDNPA, is shown in Figure 8. The total running time for a complete analysis in a microbore chromatographic system is about 80 min, including column recycling. In the HPLC profile, each D-amino acid derivative is baseline-separated from the corresponding L stereoisomer with only one coelution, D-valine with L-lysine. In most cases, each amino acid derivative can be detected and quantitated in the presence of a large excess of the corresponding diastereoisomer. All amino acid derivatives show a linear relationship between the amount injected and the peak area in the range of 0.5 to 3 nmol. The high recovery of each DNPA-amino acid (with negligible racemization) after 5 h hydrolysis in vapor phase with 6 N HCl at 110°C renders FDNPA an attractive reagent for the analysis of the chiral identity of amino acids in the course of sequence determination of peptides according to a variation of the classical dansyl-Edman degradation. The Edman chemistry is used merely as a subtractive procedure. After each cycle of degradation, a suitable aliquot of the shortened peptide is subjected to end-group analysis by reaction with FDNPA. The chirality of the DNPA-amino acid released after acid hydrolysis of the peptide may be determined by RP-HPLC as described [83]. A modification of the hydrolysis procedure with the addition of scavengers in the acid solution in order to maximize the recovery of the N-terminal DNPA residues was subsequently reported [73]. Furthermore, when 1,5-difluoro-2,4-dinitrobenzene, suspended in a few microliters of acetone, is placed in a separate tube in the vapor phase hydrolysis apparatus, the highest recoveries are observed. As the commercially available Marfey reagent (FDAA) is subjected to deamidation during acid hydrolysis, it can be used to produce DNPA-amino acids suitable for HPLC analysis and to determine the chirality of N-terminal amino acid residues, sequentially exposed in the course of subtractive Edman degradation of peptides. This procedure allowed the determination of the chiral sequence of Met-deltorphin and other synthetic peptides [14], as well as lipodepsipeptides [75]. The identification of the chirality of the DNPA-derivatives released after Edman degradation was always unequivocal and allowed us to verify that the racemization at each step of automated peptide synthesis was less than 2%. However, in the case of two synthetic peptides, Tyr-DAsp-Phe-Asp-Val-Val-GlyNH_2 and Tyr-DLys-Phe-Asp-Val-Val-GlyNH_2, a significant extent of racemization was observed. In particular, the amino acid composition of the first peptide, determined after acid hydrolysis and derivatization with FDNPA, was in accord with that expected, but the ratio L/D-Asp was found to be 0.62/1. Determination of the chiral identity of amino acid residues in the course of Edman degradation of this peptide revealed the presence of 3% L vs. D isomer of aspartic acid in position 2 and 31% of

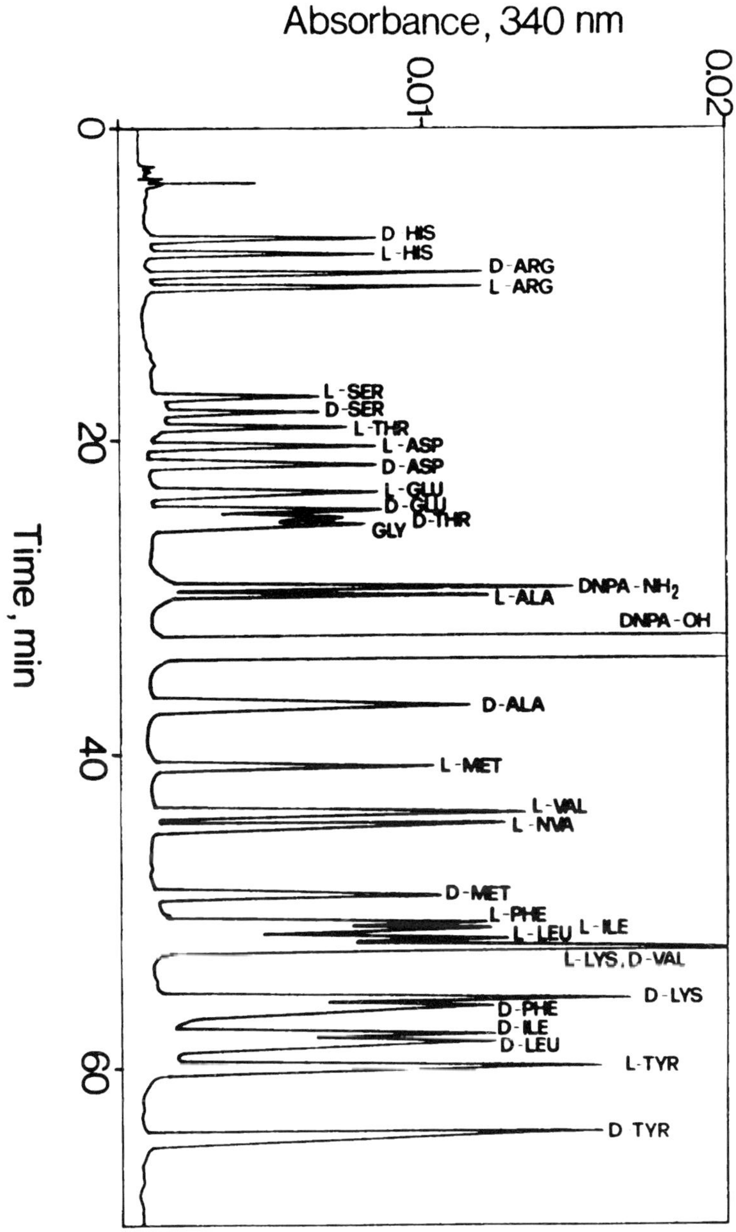

Figure 8. Separation of a racemic amino acid mixture plus glycine after derivatization with 1-fluoro-2,4-dinitrophenyl-5-L-alanine (FDNPA). Sample size, 400 pmol of each amino acid. DNPA-NH_2, 1-amino-2,4-dinitrophenyl-5-L-alanine; DNPA-OH, 1-hydroxy-2,4-dinitrophenyl-5-L-alanine; NVA, norvaline. (From ref. 84).

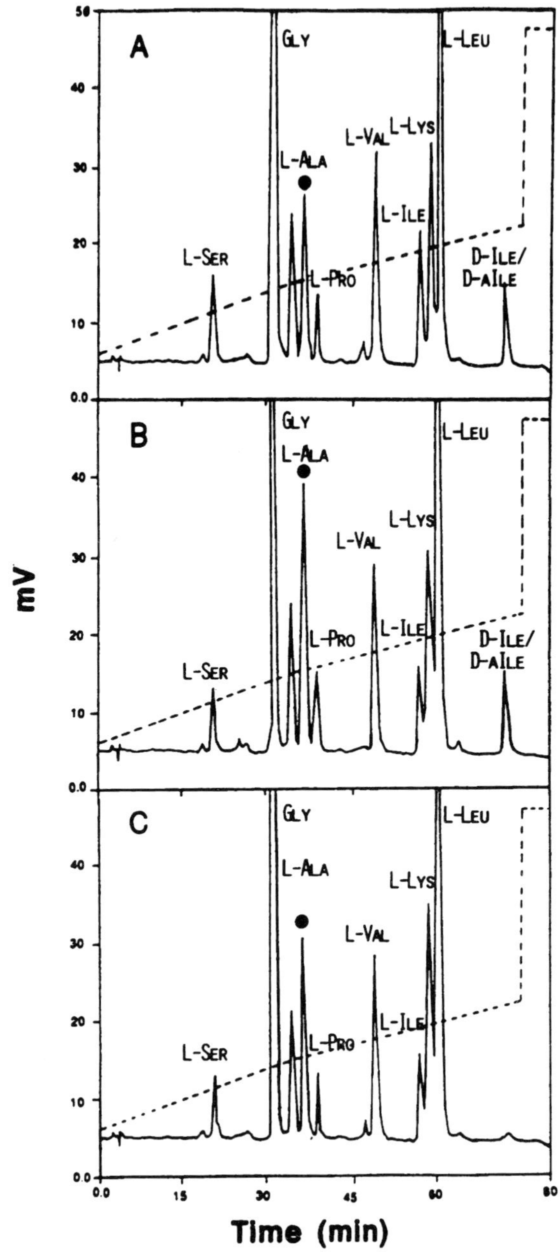

Figure 9. Chiral amino acid analysis of bombinin H5 [17] as such (A) and after the first (B) and second (C) cycle of Edman degradation. The peak denoted with • is reagent-releted.

D vs. L isomer of the same amino acid in position 4. This indicates the occurrence of some difficulties during the synthesis of this heptapeptide. Since both molecules were synthesized simultaneously for the first five residues, a similar extent of racemization was expected. In fact, the sequential analysis of the second heptapeptide revealed the presence of 32% of D vs. L isomer in position 4 [14]. The comparison of the extent of racemization observed in the course of compositional analysis with that observed during sequential degradation of peptides confirmed the effect of acid hydrolysis on this reaction [77]. With the same procedure, the complete stereochemistry of different antibacterial peptides from cultures of *Pseudomonas syringae* and *Pseudomonas fuscovaginae* has been determined [84–88]. In the case of bombinins H [17], the D-*allo*isoleucine in the second position of the sequence was identified by both ion-exchange and chiral amino acid analyses of peptide aliquots after each cycle of Edman degradation. In fact, in the chiral analysis D-*allo*Ile and D-Ile coelute (and are well separated from the L-Ile/L-*allo*Ile peak), whereas by ion-exchange chromatography *allo*Ile is well separated from Ile. Therefore, the combination of the two analyses allowed an unequivocal assignment. The results of the chiral HPLC analysis of bombinin H5 [17] after two cycles of Edman degradation are reported in Figure 9.

Perspectives: Mass spectrometry?

Racemic amino acid mixtures cannot be distinguished by mass spectrometry. Only in the case of Ile-*a*Ile and Thr-*a*Thr pairs have differences in the intensity of selected peaks present in the fragmentation pattern been observed by electron impact techniques for *N*,*O*(*S*)-*tert*-butyl dimethylsilyl derivatives (A. Scaloni, unpublished observations). Discrimination between enantiomers by using a chiral reagent has always been a challenge for mass spectrometrists. The use of chiral crown ethers as host molecules and the investigation of their adduct formation with chiral compounds have attracted a lot of interest during the last few years [89] and possibly will be a promising approach for amino acid enantiomeric resolution.

Acknowledgements
Work in the laboratories of the authors has been supported by grants from the Ministero dell' Università e della Ricerca Scientifica e Tecnologica and from the National Research Council Special Project on Structural Biology.

References

1. Kleinkauf, H. and von Dohren, H. (1990) Nonribosomal biosynthesis of peptide antibiotics. *Eur J Biochem* 192: 1–15.
2. Bada, J.L. (1984) *In vivo* racemization in mammalian proteins. *Methods Enzymol* 106: 98–115.
3. Lubec, G. and Lubec, B. (1993) Racemization rates of amino acids for dating ancient samples. *Amino Acids* 4: 1–3.

4. Nagata, Y., Akino, T., Ohno, K., Kataoka, Y., Ueda, T., Sakura, T., Shiroshita, K. and Yasuda, T. (1987) Free D-amino acids in human plasma in relation to senescence and renal diseases. *Clin Sci* 73:105–108.
5. Dunlop, D.S., Neidle, A., McHale, D.M., Dunlop, D.M. and Lajtha, A. (1986) The presence of free D-aspartic acid in rodents and man. *Biochem Biophys Res Commun* 141:27–32.
6. Nagat, Y., Yamamoto, K. and Shimojo, T. (1992) Determination of D- and L-amino acids in mouse kidney by HPLC. *J Chromatogr* 575:147–152.
7. Felbeck, H. and Wiley, S. (1987) Free D-amino acids in the marine molluscs. *Biol Bull* 173:252–259.
8. D'Aniello, A., Di Cosmo, A., Di Cristo, C. and Fisher, G. (1992) D-aspartate in the male and female reproductive system of *Octopus vulgaris*. *Gen Comp Endocrinol* 100: 69–72.
9. Bruckner, H., Jaek, P., Langer, M. and Godel, H. (1994) Liquid chromatographic determination of D-amino acids in cheese and cow milk. Implication of starter cultures, amino acid racemases and rumen microorganisms on formation, and nutritional considerations. *Amino Acids* 2:271–284.
10. Nagata, Y., Borghi, M., Fisher, G.H. and D'Aniello, A. (1995) Free D-serine concentration in normal and Alzheimer human brain. *Brain Res Bull* 38:181–183.
11. Kreil, G. (1997) D-Amino acids in animal peptides. *Annu Rev Biochem* 66:337–345.
12. Montecucchi, P.C., de Castiglione, R., Piani, S., Gozzini, L. and Erspamer, V. (1981) Amino acid composition and sequence of dermophin, a novel opiate-like peptide from the skin of *Phyllomedusa sauvagei*. *Int J Pept Prot Res* 17:275–283.
13. Richter, K., Egger, R. and Kreil, G. (1987) D-Alanine in the frog skin peptide dermorphin is derived from L-alanine in the precursor. *Science* 238:200–202.
14. Ersparmer, V., Melchiorri, P., Falconieri-Ersparmer, G., Negri, L., Corsi, R., Severini, C., Barra, D., Simmaco, M. and Kreil, G. (1989) Deltorphins, a family of naturally occurring peptides with high affinity and selectivity for delta opioid binding sites. *Proc Natl Acad Sci USA* 86:5188–5192.
15. Mignogna, G., Severini, C., Simmaco, M., Negri, L., Falconieri-Erspamer, G., Kreil G. and Barra, D. (1992) Identification and characterization of two dermorphins from skin estracts of the Amazonian frog *Phyllomedusa bicolor*. *FEBS Lett* 302:151–154.
16. Barra, D., Mignogna, G., Simmaco, M., Pucci, P., Severini, C., Falconieri-Erspamer, G., Negri, L. and Erspamer, V. (1994) [L-Leu^2]Deltorphin, a 17 amino acid opioid peptide from the skin of the Brazilian hylid frog *Phyllomedusa burmeisteri*. *Peptides* 15: 199–202.
17. Mignogna, G., Simmaco, M., Kreil, G. and Barra, D. (1993) Antibacterial and haemolytic peptides containing D-alloisoleucine from the skin of *Bombina variegata*. *EMBO J* 12: 4829–4832.
18. Meister, A. (ed.) (1965) *Biochemistry of the Amino Acids*, vol. 1. Academic Press, New York.
19. Wakayama, M., Takashima, K., Tau, Y., Nakashima, S., Sakai, K. and Moriguchi, M. (1997) Spectrophotometric assay of D-aspartate and D-glutamate using D-aspartate oxidase with malate dehydrogenase and glutamate dehydrogenase. *Anal Biochem* 250:252–253.
20. Jones, W.M., Ringe, D., Soda, K., Manning and J.M. (1994) Determination of free D-amino acids with bacterial transaminase: their depletion leads to inhbition of bacterial growth. *Anal Biochem* 218:204–209.
21. Davankov, V.A., Rogozhin, S.V., Semechkin, A.Y. and Sachkova T.P. (1973) Ligand-exchange chromatography of racemates. Influence of the degree of saturation of the asymmetric resin by metal ions on ligand exchange. *J Chromatogr* 82:359–365.
22. Gubitz, G., Jellenz, W. and Santi, W. (1981) Separation of the optical isomers of amino acids by ligand-exchange chromatography using chemically bonded chiral phases. *J Chromatogr* 203:377–384.
23. Hare, P.E. and Gil-Av, E. (1979) Separation of D and L amino acids by liquid chromatography: use of chiral eluants. *Science* 204:1226–1228.
24. Weinstein, S., Engel, M.H. and Hare, P.E. (1982) The enantiomeric analysis of a mixture of all common protein amino acids by HPLC using a new chiral mobile phase. *Anal Biochem* 121:370–377.
25. Manning, J.M., Moore, S. (1968) Determination of D- and L-amino acids by ion exchange chromatography as L-D and L-L-dipeptides. *J Biol Chem* 243:5591–5597.

26. Mitchell, A.R., Kent, S.B.H., Chu, I.C. and Merrifield, R.B. (1978) Quantitative determination of D- and L-amino acids by reaction with tert-butyloxycarbonyl-L-leucine *N*-hydroxysuccinimide ester and chromatographic separation as L,D and L,L dipeptides. *Anal Chem* 50:637–640.
27. Souter, R.W. (1985) *Chromatographic separations of stereoisomers.* CRC Press, Boca Raton.
28. Allenmark, S.G. (1988) *Chromatographic enantio-separation: methods and applications.* Ellis Horwood, Chichester.
29. Pirkle, W.H. and Pochapsky, T.C. (1989) Considerations of chiral recognition relevant to the liquid chromatographic separation of enantiomers. *Chem Rev* 89:347–362.
30. Yuasa, S., Itoh, M. and Shimada, A. (1984) Resolution of amino acids by a native-cellulose column. *J Chromatogr* 22:288–292.
31. Blaschke, G. and Maibaum, J. (1986) Chromatographic resolution. Optical resolution of the racemic anticancer drug ifosfamide and other chiral oxazaphosphorines. *J Chromatogr* 366:329–334.
32. Allenmark, S. (1993) Chiral discrimination by albumin. *Chirality* 5:295–299.
33. Jadaud, P., Thelohan, S., Schonbaum, G.R. and Wainer, I.W. (1989) The stereochemical resolution of enantiomeric free and derivatized amino acids using an HPLC chiral stationary phase based on immobilized alpha-chymotrypsin: chiral separation due to solute structure or enzyme activity. *Chirality* 1:38–44.
34. Frank, H. (1988) Chiral stationary phases for capillary gas chromatography. *J High Resol Chromat Commun* 11:787–792.
35. Ketting, D., Wadman, S.K., Spaapen L.J.M., Van der Meer, S.B. and Duran, M. (1991) Gas chromatography method for the separation of amino acids enantiomers in plasma and urine. Application in a case of short bowel syndrome. *Clin Chim Acta* 204:79–86.
36. Hara, S. and Dobashi, A. (1979) Liquid chromatographic resolution of enantiomers on chiral amide bonded silica gel normal phase separation of racemica alpha amino-acid derivatives by N-acetyl-L-valylaminopropyl silanized silica phase. *J Liq Chromatogr* 2:883–890.
37. Pirkle, W.H. (1983) Stacking of amide dipoles as an alternative to hydrogen bonds in chiral recognition rationales. *Tetrahedron Lett* 24:5707–5708.
38. Fukushima, T., Kato, M., Santa, T. and Imai, K. (1995) Enantiomeric separation and sensitive determination of D,L-amino acids derivatized with fluorogenic benzofuran reagents on Pirkle type stationary phases. *Biomed Chrom* 9:10–17.
39. Imai, K. and Fukushima, T. (1993) Derivatization with fluorogenic benzofuran reagents of amino acid enantiomers and their separation on a Pirkle type column. *Biomed Chrom* 7:273–281.
40. Imai, K., Fukushima, T. and Uzu, S. (1993) Sensitive determination of enantiomers of amino acids derivatized with the fluorogenic reagent, 4-fluoro-7-nitro-2,1,3-benzoxadiazole, separated on a Pirkle-type column, Sumichiral OA 2500. *Biomed Chrom* 7:177–181.
41. Pirkle, W.H., Finn, J.M., Schreiner, J.L. and Hamper, B.C. (1981) A widely useful chiral stationary phase for HPLC separation of enantiomers. *J Am Chem Soc* 103:3764–3794.
42. Armstrong, D.W., De Mond, W. and Czech, B.P. (1985) Separation of metallocene enantiomers by liquid chromatography: chiral recognition via cyclodextrin bonded phases. *Anal Chem* 57:481–484.
43. Nimura, N., Suzuki, T., Kasahara, Y. and Kinoshita, T. (1981) Reversed-phase liquid chromatographic resolution of amino acid enantiomers by mixed chelate complexation. *Anal Chem* 53:1380:1383.
44. Lam, S. (1984) Stereoselective analysis of D and L dansyl amino acids as mixed chelate copper(II) complexes by HPLC. *J Chromatogr Sci* 22:416–423.
45. Armani, E., Dossena, A., Marchelli, R. and Virgili, R. (1988) Copper(II) complexes of di-amino-diamido-type ligands as chiral eluents in the enantiomeric separation of D,L-dansylamino acids by RP-HPLC. *J Chromatogr* 441:275–286.
46. Armani, E., Barazzoni, L., Dossena, A. and Marchelli, R. (1988) Bis(L-amino acid amidato) copper(II) complexes as chiral eluents in the enantiomeric separation of D,L-dansylamino acids by RP-HPLC. *J Chromatogr* 441:287–298.
47. Belov, Y., Aksinenko, A., Blessington, B. and Newman, A.H. (1996) Chiral aminophosphonate eluent for enantiomeric analysis of amino acids. *Chirality* 8:122–125.

48. Van der Haar, J., Kip, J. and Kraak, J.C. (1988) Effect of the mobile phase composition and ligand structure on the separation of D- and L-dansylamino acids, as mixed metal complexes, by reversed-phase high performance liquid chromatography. *J Chromatogr* 445: 219–224.
49. Armstrong, D.W. and Han, S.M. (1987) Use of microcolumn liquid chromatography with a chiral stationary phase for the separation of low-resolution enantiomers. *J Chromatogr* 389:256–260.
50. Riester, D., Wiesmuller, K., Stoll, D. and Kuhn, R. (1996) Racemization of amino acids in solid-phase peptide synthesis investigated by capillary electrophoresis. *Anal Chem* 68: 2361–2365
51. Copper, C.L., Davis, J.B., Cole, R.O. and Sepaniak, M.J. (1994) Separation of derivatized amino acid enantiomers by cyclodextrin-modified capillary electrophoresis: mechanistic and molecular modeling studies. *Electrophoresis* 15:785–792.
52. Florance, J., Galdes, A., Konteatis, Z., Kosarych, Z., Langer, K. and Martucci, C. (1987) High performance liquid chromatographic separation of peptide and amino acid stereoisomers. *J Chromatogr* 414:313–322.
53. Pawlowska, M., Chen, S. and Armstrong, D.W. (1993) Enantiomeric separation of fluorescent, 6-aminoquinolyl-*N*-hydroxysuccinimidyl carbamate, tagged amino acids. *J Chromatogr* 641:257–265.
54. Furukawa, H., Sakakibara, E., Kamei, A. and Ito, K. (1975) Separation of L- and D-amino acids as diastereomeric derivatives by high performance liquid chromatography. *Chem Pharm Bull* 23:1625–1626.
55. Goto, J., Hasegawa, M., Nakamura, S., Shimada, K. and Nambara, T. (1978) New derivatization reagents for the resolution of amino acid enantiomers by high performance liquid chromatography. *J Chromatogr* 152:413–419.
56. Nambara, T., Ikegawa, S., Hasegawa, M. and Goto, J. (1978) High pressure liquid chromatographic resolution of amino acid enantiomers by derivatization with new chiral reagents. *Anal Chim Acta* 101:111–116.
57. Nimura, N., Ogura, H. and Kinoshita, T. (1980) Reversed-phase liquid chromatographic resolution of amino acid enantiomers by derivatization with 2,3,4,6-tetra-*O*-acetyl-β-D-glucopyranosyl isothiocyanate. *J Chromatogr* 202:375–379.
58. Kinoshita, T., Kasahara, Y. and Nimura, N. (1981) Reversed-phase high performance liquid chromatographic resolution of nonesterified enantiomeric amino acids by derivatization with 2,3,4,6-tetra-*O*-acetyl-β-D-glucopyranosyl isothiocyanate and 2,3,4-tri-*O*-acetyl-α-arabinopyranosyl isothiocyanate. *J Chromatogr* 210:77–81.
59. Nimura, N., Toyama, A. and Kinoshita, T. (1984) Optical resolution of amino acid enantiomers by HPLC. *J Chromatogr* 316:547–552.
60. Davies, J.S. and Mohammed, A.K.A. (1984) Chiral analysis of the reaction stages in the Edman method for sequencing peptides. *J Chem Soc Perkin Trans II*: 1723–1727.
61. Dunlop, D.S. and Neidle, A. (1987) The separation of D/L amino acid pairs by high-performance liquid chromatography after precolumn derivatization with optically active naphthylethyl isocyanate. *Anal Biochem* 165:38–44.
62. Aswad, D.W. (1984) Determination of D and L-aspartate in amino acid mixtures by high performance liquid chromatography after derivatization with a chiral adduct of *o*-phthaldialdehyde. *Anal Biochem* 137:405–409.
63. Buck, R.H. and Krummen, K. (1984) Resolution of amino acid enantiomers by high performance liquid chromatography using automated pre-column derivatization with a chiral reagent. *J Chromatogr* 315:279–285.
64. Nimura, N. and Kinoshita, T. (1986) *O*-Phthalaldehyde-*N*-acetyl-L-cysteine as a chiral derivatization reagent for liquid chromatographic optical resolution of amino acid enantiomers and its application to conventional amino acid analysis. *J Chromatogr* 352: 169–177.
65. Jegorov, A., Triska, J., Trnka, T. and Cerny, M. (1988) Separation of alfa amino acid enantiomers by reversed phase high performance liquid chromatography after derivatization with *o*-phthaldialdehyde and a sodium salt of 1-thio-β-D-glucose. *J Chromatogr* 434: 417–422.
66. Bruckner, H., Haasmann, S., Langer, M., Westhauser, T., Wittner, R. and Godel, H. (1994) Liquid chromatographic determination of D and L-amino acids by derivatization with *o*-phthaldialdehyde and chiral thiols. *J Chromatogr* 666:259–273.

67. Einarsson, S., Josefsson, B., Moller, P. and Sanchez, D. (1987) Separation of amino acid enantiomers and chiral amines using precolumn derivatization with 1-(9-fluorenyl)ethyl-chloroformate and reversed phase liquid chromatography. *Anal Chem* 59:1191–1195.
68. Einarrson, S., Josefsson, B. and Lagerkvist, S. (1983) Determination of amino acids with 9-fluorenylmethyl chloroformate and reversed phase high performance liquid chromatography. *J Chromatogr* 282:609–618.
69. Chan, K.C., Muschik, G.M. and Issaq H.J. (1995) Enantiomeric separation of amino acids using micellar electrokinetic chromatography after pre-column derivatization with the chiral reagent 1-(9-fluorenyl)-ethyl chloroformate. *Electrophoresis* 16:504–509.
70. Marfey, P. (1984) Determination of D-amino acids. II. Use of a bifunctional reagent, 1,5-difluoro-2,4-dinitrobenzene. *Carlsberg Res Commun* 49:591–596.
71. Szokan, G., Mezo, G. and Hudecz, F. (1988) Application of Marfey's reagent in racemization studies of amino acids and peptides. *J Chomatogr* 444:115–122.
72. Kochhar, S. and Christen, P. (1989) Amino acid analysis by HPLC after derivatization with 1-fluoro-2,4-dinitrophenyl-5-L-alanine amide. *Anal Biochem* 178:17–21.
73. Scaloni, A., Simmaco, M. and Bossa, F. (1995) D-L amino acid analysis using automated precolumn derivatization with 1-fluoro-2,4-dinitrophenyl-5-alanine amide. *Amino Acids* 8:305–313.
74. Bruckner, H. and Keller-Hoehl, C. (1990) HPLC separation of DL-amino acids derivatized with N^2-(5-fluoro-2,4-dinitrophenyl)-L-amino acid amides. *Chromatographia* 30:621–629.
75. Bachmann, C., Ballio, A., Barra, D., Bossa, F., Camoni, L., Di Giorgio, D., Di Nola, A., Ferranti, P., Grgurina, I. and Iacobellis, N.S. et al. (1996) Lipodepsipeptides from *Pseudomonas syringae* pv *syringae*. In: K. Rudolph, T.J. Burr, J.W. Mansfield, D. Stead, V. Vivian, J. von Kietzel (eds): *Pseudomonas syringae pathovars and related pathogens*. Kluver Academic Publishers, Dordecht, The Netherlands, 170–174.
76. Tran, A.D., Blanc, T. and Leopold, E.J. (1990) Free solution capillary electrophoresis and micellar electrokinetic resolution of amino acid enantiomers and peptide isomers with L- and D-Marfey's reagents. *J Chromatogr* 516:241–249.
77. Neuberger, A. (1948) Stereochemistry of amino acids. *Adv Prot Chem* 4:297–383.
78. Manning, J.M. (1970) Determination of D- and L-amino acid residues in peptides. Use of tritiated hydrochlorid acid to correct for racemization during acid hydrolysis. *J Am Chem Soc* 92:7449:7454.
79. Goodlett, D.R., Abuaf, P.A., Savage, P.A., Kowalski, K.A., Mukherjee, T.K., Tolan, J.W., Corkum, N., Goldstein, G. and Crowther, J.B. (1995) Peptide chiral purity determination: hydrolysis in deuterated acid, derivatization with Marfey's reagent and analysis using high-performance liquid chromatography-electrospray ionization-mass spectrometry. *J Chromatogr* 707:233–244.
80. Edman, P. (1950) Preparation of phenyl thiohydantoins from some natural amino acids. *Acta Chem Scand* 4:277–282.
81. Imai, K., Matsunaga, H., Fukushima, T., Santa, T., Homma, H., Nakashima, K. and Akiyama, S. (1995) A novel method for the amino acid sequence/configuration determination of peptides containing D/L-amino acids utilizing a fluorogenic Edman reagent, 7-*N,N*-dimethylaminosulphonyl-4-(2,1,3-benzoxadiazolyl)isothiocyanate (DBD-NCS). *Biomed Chromatogr* 9:152–154.
82. Bolte, T., Yu, D., Stuwe, H.T. and Konig, W.A. (1987) Sequential configuration analysis of peptides. *Angew Chem Int Ed Engl* 4:331–332.
83. Scaloni, A., Simmaco, M. and Bossa, F. (1991) Determination of the chirality of amino acid residues in the course of subtractive Edman degradation of peptides. *Anal Biochem* 197:305–310.
84. Ballio, A., Bossa, F., Di Giorgio, D., Ferranti, P., Paci, M., Pucci, P., Scaloni, A., Segre, A. and Strobel, G.A. (1994) Novel bioactive lipodepsipeptides form *Pseudomonas syringae*. *FEBS Lett* 355:96–100.
85. Ballio, A., Bossa, F., Di Giorgio, D., Di Nola, A., Manetti, C., Paci, M., Scaloni, A. and Segre A. (1995) Solution conformation of the *Pseudomonas siringae* pv *syringae* phytotoxic lipodepsipeptide syringopeptin 25-A. *Eur J Biochem* 234:747–758.
86. Ballio, A., Bossa, F., Camoni, L., Di Giorgio, D., Flamand M.C., Maraite, H., Nitti, G., Pucci, P. and Scaloni, A. (1996) Structure of fuscopeptins, phytotoxic metabolites from *Pseudomonas fuscovaginae*. *FEBS Lett* 381:213–216.

87. Scaloni, A., Bachmann, R.C., Takemoto, J.Y., Barra, D., Simmaco, M. and Ballio, A. (1994) Stereochemical structure of Syringomycin, a phytotoxic metabolite of *Pseudomonas syringae* pv *syringae. Natur Product Lett* 4: 159–164.
88. Ballio, A., Bossa, F., Collina, A., Gallo, M., Iacobellis, N.S., Paci, M., Pucci, P., Scaloni, A., Segre, A. and Simmaco, M. (1990) Structure of syringotoxin, a bioactive metabolite of *Pseudomonas syringae* pv *syringae. FEBS Lett* 269:377–380.
89. Pocsfalvi, G., Liptak, M., Huszthy, P., Bradshaw, J.S., Izatt, R.M. and Vekey, K. (1992) Characterization of chiral host-guest complexation in fast atom bombardment mass spectrometry. *Anal Chem* 68:792–795.
90. Cabrera, K. and Schwinn, G. (1990) Cyclodextrins as mobile phase additives in the HPLC separation of enantiomers. *Int Laboratory*, Jul–Aug, 28–32.

Occurrence and function of D-amino acids-containing peptides and proteins

D-Amino Acids in Sequences of Secreted Peptides of Multicellular Organisms
ed. by P. Jollès

Occurrence and function of D-amino acid-containing peptides and proteins: Antimicrobial peptides

G. Mignogna [1], M. Simmaco [2] and D. Barra [1]

[1] *Istituto Pasteur-Fondazione Cenci Bolognetti, Dipartimento di Scienze Biochimiche "A. Rossi Fanelli" and CNR Centro di Biologia Molecolare, Università La Sapienza, Piazzale Aldo Moro 5, I-00185 Roma, Italy*

[2] *Dipartimento di Scienze Biomediche, Università di Chieti G. D'Annunzio, Via dei Vestini, I-66100 Chieti, Italy*

Summary. Antimicrobial peptides are widely distributed in living organisms, where they represent a constitutive defence system acting as a first line of response against infections. The number of such peptides discovered has increased rapidly in the last few years, and more than 100 have been described from different sources. So far, antimicrobial peptides containing a D-amino acid have only been found in the skin secretions of frogs belonging to the genus *Bombina*. In the second position of the sequence of the mature peptides either D-alloisoleucine or D-leucine were detected. The D-amino acids are derived from the corresponding L forms by an as yet unknown posttranslational reaction.

Introduction

Antimicrobial peptides are widely distributed in living organisms, where they represent a constitutive defence system acting as a first line of response against infections. They are generally recognized as effector molecules in innate immunity [1], as they confer protection against pathogens not only in lower organisms but also in vertebrates, which are additionally provided with an extremely complex and efficient adaptive immune system. The number of such peptides discovered has increased rapidly in the last few years, and more than 100 have been described from different sources (cells, tissues, fluids, glands of different animals). A database of antimicrobial peptides is available at the Web site http://www.univ.trieste.it/nirdbbcm/tossi/. The antibacterial peptides have been grouped according to their structural characteristics: (i) linear peptides without cysteines, mostly helical, like cecropins from insects [2], magainins [3] and PGLa [4], bombinins [5, 6] and bombinins H [7], dermaseptins [8] and temporins [9] from amphibians (for a review, see ref. 10); (ii) linear peptides without cysteines, but showing an overrepresentation of certain amino acids, like Pro and Arg as seen in honeybee apidaecin [11], bovine bactenecins [12], porcine PR-39 [13] and Trp in indolicidin from neutrophils [14]; (iii) peptides containing one disulfide bridge, like the brevinins and esculentins from *Rana* species [10, 15]; and (iv) peptides with two or more disulfide bridges, like defensins from mammalian neutrophils [16], tachyplesins from crustaceans [17] and NK-lysin from pig intestine [18]. Despite the high structural variability and differing lengths (ranging from the 13 amino acid residues of temporins [9] to the 46 residues of esculentins [15]), all antimicrobial peptides are positively charged, mostly amphipathic and the proposed mechanism of action is mainly based on the selective permeabilization of bacterial cell membranes, by interaction with their anionic phospholipids.

Most classical antibiotics are made by microbes through stepwise enzymatic assembly of several components, whereas antimicrobial peptides from higher organisms are all gene-encoded. They are synthesized on ribosomes as preproproteins, and subsequently proteolytically processed in several steps to produce the active forms. In some cases they are subjected to further posttranslational modifications. The formation of one or more disulfide bridges has already been mentioned as typical of some families of antimicrobial peptides; amidation of the C-terminal extremity has been frequently observed. Moreover, modification of the N-terminus or glycosylation has occasionally been found [19]. A specific role for such modifications has been assigned only in few cases, e.g., for achieving the proper, biologically active conformation, as in the case of disulfide bridges of defensins, or for the recognition of specific structural components of the target cell membrane. Modifications of the extremities of the peptide chains are considered a device for controlling the proteolytic susceptibility of these molecules, thus playing an important role in their biostability.

Antimicrobial peptides from *Bombina variegata*

A peculiar case is represented by the cytolytic peptides found in skin secretions of frogs belonging to the genus *Bombina*. The occurrence of potent antimicrobial and hemolytic activities in the skin secretion of *B. variegata* has been known for a long time [20]. These activities were ascribed to a 24-residue peptide termed "bombinin" [21]. More recently, a peptide fraction was isolated from methanolic skin extracts of this frog that showed antibacterial but no hemolytic properties [6]. The fraction contained several peptides related to bombinin, but differing from each other by only one or a few amino acids. These peptides displayed a variable amino-terminal sequence and an identical carboxy-terminal region. They all have a C-terminal amide and contain 27 amino acids, 3 more than bombinin [21]. A complementary DNA (cDNA) library prepared from skin of *B. variegata* was screened with synthetic oligonucleotides derived from the constant carboxy-terminal region, and several clones were isolated which encoded the precursors of the different bombinin-like peptides [6]. Analysis of the amino acid sequence of five of these precursors revealed that each of them contained an additional message for the synthesis of another peptide, highly hydrophobic, identical in all clones. Since the crude secretion from *Bombina* was both antimicrobial and strongly hemolytic, whereas the peptides isolated from methanolic skin extracts were antimicrobial but not hemolytic, attention was focused on the hydrophobic peptide deduced from the cDNA. It has a predicted amino-terminal sequence Ile-Ile-Gly... and, depending on the processing of the precursor polypeptide, a carboxy-terminal sequence Leu-Leu or Leu-Leu-Lys-Lys-Ile-amide, yielding a 17- or 20-residue peptide, respectively [6]. These two hypothetical peptides were synthesized and used as markers to search for their natural counterparts. In this study, the skin secretion of *B. variegata*, rather than the methanolic extract, was used, obtained by discharging the glands via a mild surface electrical stimulation, a procedure that was shown to be well tolerated by the animal. When the crude skin secretion from *B. variegata* was fractionated by reversed-phase HPLC (RP-HPLC), using suitably modified elution conditions, five peptide peaks were detected, well separated from the bombinin-like peptides and eluting in the same region as the synthetic peptide terminating with Ile-amide [7]. These peptides were termed bombinin H (H1–H5), since they are *h*ydrophobic and have both antibacterial and *h*emolytic activities. Thus, differences in the procedure for collection of the starting material (mild electrical stimulation vs. methanolic extract), exploitation of the high resolving capacity of RP-HPLC techniques allowed identification of the factors responsible for the hemolytic activity of *B. variegata* skin secretion.

The structure of bombinin H was determined by Edman degradation. The peptide fractions corresponding to the five peaks were shown to differ in only two positions (position 1, occupied by either isoleucine or leucine, and position 8, by methionine or leucine), thus confirming a lower extent of variability as compared with the bombinins. However, further analysis by ion-exchange amino acid analysis revealed that three of the peptides contained up to 0.8 mol of *allo*isoleucine per mol/l of peptide, a quantity far in excess of the trace amounts usually present in protein acid hydrolysates [7]. As the peptide was gene-encoded, a posttranslational reaction was hypothesized involving racemization of a single isoleucine residue to alloisoleucine. Since several amphibian opioid peptides, such as dermorphins [22] and deltorphins [23], are known to contain D-amino acids in the second position of their sequence, the occurrence of a D-amino acid in bombinins H was suspected. In fact, supposing a similar mechanism as in opiates, the change of configuration at the α-carbon atom, leaving unchanged the β-one, should result in the conversion of L-isoleucine to D-alloisoleucine (Fig. 1). Subsequently, this residue was identified in acid hydrolysates of bombinins H3–H5 following chiral amino acid analysis [7]. Chiral subtractive Edman degradation [24] of these peptides confirmed the presence of D-alloisoleucine in the second position, whereas all other

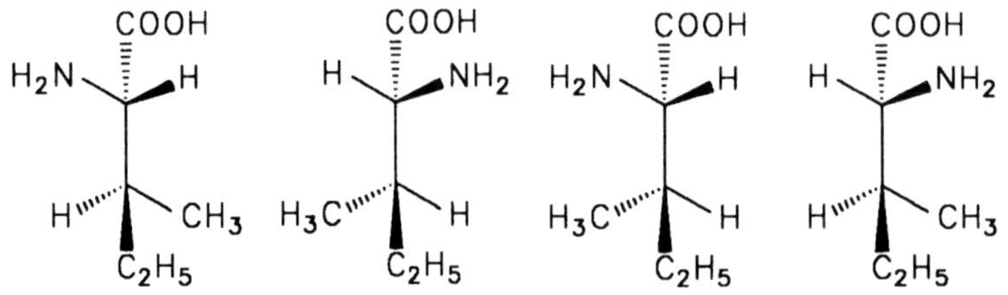

L-Ile D-Ile L-aIle D-aIle

Figure 1. Isoleucine stereoisomers.

Bombinin H1	IIGPVLGMVGSALGGLLKKIa
Bombinin H2	LIGPVLGLVGSALGGLLKKIa
Bombinin H3	I**I**GPVLGMVGSALGGLLKKIa
Bombinin H4	L**I**GPVLGLVGSALGGLLKKIa
Bombinin H5	I**I**GPVLGLVGSALGGLLKKIa
L-17-residue peptide	ILGPILGLVSNALGGLLa
D-17-residue peptide	I**L**GPILGLVSNALGGLLa

Figure 2. Sequences of bombinins H and the 17-residue peptides. D-amino acids are boldfaced. **I**, alloisoleucine; a, amide.

residues in the sequence, including isoleucine in positions 1 and 20, retained the L configuration. It should be noted that, although in RP-HPLC chiral analysis [24] D-isoleucine and D-alloisoleucine coelute, their peak is well separated from that of the corresponding L isomers.

The sequences of bombinins H are reported in Figure 2. The first eluting components (H1–H2) contain L-isoleucine in the second position, whereas the later eluting peaks (H3–H5) contain D-alloisoleucine. This is in line with earlier observations on dermorphins and deltorphins, where the D-amino acid-containing peptides eluted later than the corresponding L isomer, under similar experimental conditions of RP-HPLC.

Peptides from *B. orientalis*

Gibson et al. described antimicrobial peptides similar to bombinin in *B. orientalis* [5]. From the sequences of clones obtained from a cDNA library of dorsal skin, two copies of bombinin-like peptides were present in each clone, and the sequence of another peptide, termed C-terminal peptide, was predicted. The corresponding 17-residue amidated peptide was identified at relatively low abundance in the skin secretions by mass spectrometry. Its sequence (Fig. 2) is similar to bombinins H, but lacks the basic residues at the C-terminal extremity, which is amidated. The presence of a D-amino acid in this peptide was subsequently suggested [25]. The analysis of skin secretions from *B. orientalis*, performed in our laboratory, resulted in the identification of bombinin H-type peptides similar to those found in *B. variegata*, containing a D-alloisoleucine in the second position of the sequence (unpublished results). In addition, 17-residue peptides, identical to the C-terminal peptide predicted by Gibson [5], were found in the skin secretion, one of them with a D-Leu in the second position. The same array of molecules are present in the secretion of another frog belonging to this genus, *B. bombina* (unpublished results).

Antibacterial activity

The finding in frog skin secretion of biologically active peptides containing a D-amino acid in the second position of their sequences parallels the well-known case of the opioid peptides, dermorphins and deltorphins [22, 23], isolated from the skin secretions of a number of South American frogs. These compounds contain a D-amino acid in the same position, and a version of the peptides with an L-amino acid in the second position has not been found in the skin secretions. Moreover, synthetic all-L versions of dermorphins and deltorphins are completely inactive, the presence of the D-amino acid in the second position being essential for biological activity. These findings are opposite to the results observed with bombinins H. In fact, in the latter case, the L-isomers are present in the secretion, although in small amounts. This may reflect the lower resistence of the L isomers to the action of aminopeptidases compared with the D-amino acid-containing peptides [7].

The antibacterial activity of bombinins H was dertermined using an inhibition zone assay [26] against both the laboratory strains *Escherichia coli* D21 and *Staphylococcus aureus* CowanI and a bacterial strain isolated from the natural flora of the frog, *Aeromonas hydrophila* Bo-3N. The results indicate that all bombinins H, containing either D- or L-amino acids, display antibacterial activity of comparable efficacy (unpublished results). However, when the activity of bombinins H against *A. hydrophila* Bo-3N was tested using a liquid-based assay, the rate of the growth inhibition exerted by the D-amino acid-containing peptides was higher than the corresponding L isomers (Fig. 3).

As far as the 17-residue peptides are concerned, in the agar-based assay, they were completely inactive, possibly because of the absence of basic residues. However, the activity against *A. hydrophila* Bo-3N using the liquid-based assay shows that, while the L-Leu containing peptide is vir-

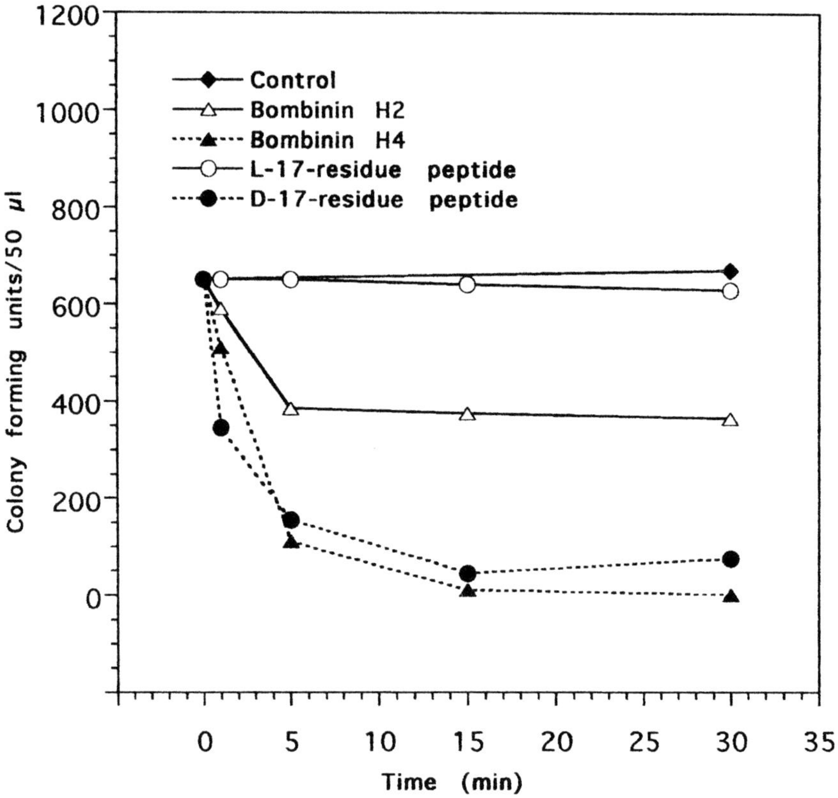

Figure 3. Rate of growth inhibition of *Aeromonas hydrophila* Bo-3N by the D and L forms of bombinin H and 17-residue peptide (20 µM).

tually inactive, the D isomer is as active as the D-amino acid-containing bombinins H (Fig. 3).

These results suggest that each peptide present in the skin secretions of *B. orientalis* or *B. bombina* has a specific function. A careful characterization of the natural bacterial flora of the animal and the analysis in suitable assay conditions of the distinctive susceptibility of the various bacterial strains to individual antibacterial peptides could explain the physiological role of this large variety of similar compounds in the skin secretions of amphibians.

Acknowledgements
Work in the laboratories of the authors has been supported by grants from the Ministero dell' Università e della Ricerca Scientifica e Tecnologica and from the National Research Council Special Project on Structural Biology.

References

1. Boman, H.G. (1995) Peptide antibiotics and their role in innate immunity. *Annuv Rev Immunol* 13:61–92.
2. Steier, H., Hultmark, D., Engström, A., Bennich, H. and Boman, H.G. (1981) Sequence and specificity of two antibacterial proteins involved in insect immunity. *Nature* 292:246–248.
3. Zasloff, M. (1987) Magainins, a class of antimicrobial peptides from *Xenopus* skin: isolation, characterization of two active forms and partial cDNA sequence of a precursor. *Proc Natl Acad Sci USA* 84:5449–5453.
4. Hoffman, W., Richter, K. and Kreil, G. (1983) A novel peptide designated PYL[a] and its precursor as predicted from cloned mRNA of *Xenopus laevis* skin. *EMBO J* 2:711–714.
5. Gibson, B.W., Tang, D., Mandrell, R., Kelly, M. and Spindel, E.R. (1991) Bombinin-like peptides with antimicrobial activity from skin secretions of the asian toad, *Bombina orientalis*. *J Biol Chem* 266:23103–23111.
6. Simmaco, M., Barra, D., Chiarini, F., Noviello, L., Melchiorri, P., Kreil, G. and Richter, K. (1991) A family of bombinin-related peptides from the skin of *Bombina variegata*. *Eur J Biochem* 199:217–222.
7. Mignogna, G., Simmaco, M., Kreil, G. and Barra, D. (1993) Antibacterial and haemolytic peptides containing D-alloisoleucine from the skin of *Bombina variegata*. *EMBO J* 12: 4829–4832.
8. Mor, A., Nguyen, V.H., Delfour, A., Migliore-Samour, D. and Nicolas, P. (1991) Isolation, amino acid sequence and synthesis of dermaseptin, a novel antimicrobial peptide of amphibian skin. *Biochemistry* 30:8824–8830.
9. Simmaco, M., Mignogna, G., Canofeni, S., Miele, R., Mangoni, M.L. and Barra, D. (1996) Temporins, antimicrobial peptides from the European red frog *Rana temporaria*. *Eur J Biochem* 242:788–792.
10. Barra, D. and Simmaco, M. (1995) Amphibian skin: a promising resource for antimicrobial peptides. *Trends Biotechnol* 13:205:209.
11. Casteels, P., Ampe, C., Jacobs, F., Vaeck, M. and Tempst, P. (1989) Apidaecins: antibacterial peptides from honeybees. *EMBO J* 8:2387–2391.
12. Frank, R.W., Gennaro, R., Schneider, K., Przybylski, M. and Romeo, D. (1990) Amino acid sequences of 2 proline-rich bactenecins, antimicrobial peptides of bovine neutrophils. *J Biol Chem* 265:18871–18874.
13. Agerberth, B., Lee, J.Y., Bergman, T., Carlquist, M., Boman, H.G., Mutt, V. and Jörnvall, H. (1991) Amino acid sequence of PR-39. Isolation from pig intestine of a new member of the family of proline-arginine-rich antibacterial peptides. *Eur J Biochem* 202:849–854.
14. Selsted, M.E., Novotny, M.J., Morris, W.L., Tang, Y.Q., Smith, W. and Cullor, J.S. (1992) Indolicidin, a novel bactericidal tridecapeptide amide from neutrophils. *J Biol Chem* 267:4292–4295.
15. Simmaco, M., Mignogna, G., Barra, D. and Bossa, F. (1994) Antimicrobial peptides from skin secretion of *Rana esculenta*. Molecular cloning of cDNAs encoding esculentin and brevinins and isolation of new active peptides. *J Biol Chem* 269:11956–11961.
16. Lehrer, R.I., Lichtenstein, A.K. and Ganz, T. (1993) Defensins, antimicrobial and cytotoxic peptides of mammalian cells. *Annu Rev Immunol* 11:105–128.
17. Nakamura, T., Furunaka, H., Miyata, T., Tokunaga, F., Muta, T., Iuanaga, S., Iwanaga, S., Niwa, M., Takao, T. and Shimonishi, Y. (1988) Tachyplesin, a class of antimicrobial peptide from hemocytes of the horseshoe crab (*Tachypleus tridentatus*). *J Biol Chem* 263: 16709–16713.
18. Andersson, M., Gunne, H., Agerberth, B., Boman, A., Bergman, T., Sillard, R., Jörnvall, H., Mutt, V., Olsson, B. and Witgzell, H. et al. (1995) NK-lysin, a novel effector peptide of cytotoxic T and NK cells. Structure and cDNA cloning of the porcine form, induction by interleukin 2, antibacterial and antitumour activity. *EMBO J* 14:1615–1625.
19. Bulet, P., Dimarcq, J.L., Hetru, C., Lagueux, M., Charlet, M., Hegy, G., Van Dorsselaer A. and Hoffmann, J.A. (1993) A novel inducible antibacterial peptide of *Drosophila* carries an O-glycosylated substitution. *J Biol Chem* 268:14893–14897.
20. Kiss, G. and Michl, H. (1962) On the venomous skin secretion of the orange speckled frog, *Bombina variegata*. *Toxicon* 1:33–30.
21. Csordas, A. and Michl, H. (1970) Isolation and structure of a haemolytic polypeptide from the defensive secretion of European *Bombina* species. *Monatsh Chemie* 101:182–189.

22. Montecucchi, P.C., de Castiglione, R., Piani, S., Gozzini, L. and Erspamer, V. (1981) Amino acid composition and sequence of dermorphin, a novel opiate-like peptide from the skin of *Phyllomedusa sauvagei. Int J Pept Prot Res* 17:275–283.
23. Erspamer, V., Melchiorri, P., Falconieri-Erspamer, G., Negri, L., Corsi, R., Severini, C., Barra, D., Simmaco, M. and Kreil, G. (1989) Deltorphins: a family of naturally occurring peptides with high affinity and selectivity for δ opioid binding sites. *Proc Natl Acad Sci USA* 86:5188–5192.
24. Scaloni, A., Simmaco, M. and Bossa, F. (1991) Determination of the chirality of amino acid residues in the course of subtractive Edman degradation of peptides. *Anal Biochem* 197: 305–310.
25. Gibson, B.W., Tang, D., Phillips, N., Medzihradsky, D., Tullius, M., Mandrell, R., Kelly, M. and Spindel, E.R. (1992) *Proceedings of the 40th ASMS Conference on Mass Spectrometry and Allied Topics.* Washington, DC 1543–1544.
26. Hultmark, D., Engström, A., Andersson, K., Steiner, H., Bennich, H. and Boman, H.G. (1983) Insect immunity. Attacins, a family of antibacterial proteins from *Hyalophora cecropia. EMBO J* 2:571–576.

D-Amino Acids in Sequences of Secreted Peptides of Multicellular Organisms
ed. by P. Jollès

Molluscan neuropeptides

Y. Yasuda-Kamatani

Suntory Institute for Bioorganic Research, 1-1-1 Wakayamadai, Shimamoto-cho, Mishima-gun, Osaka 618-8503, Japan

Summary. Achatin-I, fulicin, fulyal, *Mytilus*-FFRFamide and *Helix* CCAP-RP-III are D-amino acid-containing neuropeotides from molluscs. Achatin-I, fulicin and fulyal from *Achatina* showed excitatory and/or modulatory actions on the penis retractor, radula retractor or ventricular muscles and neurons, though their L isomers were devoid of activity. On the other hand, both *Mytilus*-FFRFamide and its L isomer showed excitatory effects on the anterior byssus retractor muscle. Moreover, in contrast to *Achatina* neuropeptides, *Helix* CCAP-RP-III exhibited no remarkable activities on any of the muscles tested; instead, its L isomer possessed various excitatory effects. The molecular structures of these short peptides would be affected by the L → D conversion and could influence activity. Molecular biological studies on the fulicin precursor suggest that fulicin, fulyal and related peptides are produced in *Achatina* ganglia and heart by processing of the ribosomally made precursor, and that L-isomeric fulicin and fulyal further undergo epimerization to yield the D-isomers.

Introduction

In molluscs, the identification of D-amino acids in sequences of peptides began in 1989 [1]. A novel neuropeptide has been isolated from the African giant snail (*Achatina fulica*) and named achatin-I from the scientific name of the snail. Until now, this peptide and four other peptides have been reported as the neuropeptides containing D-amino acids from the molluscs, i.e., achatin-I [1–3], fulicin [4, 5] and fulyal [5] from *A. fulica*, *Mytilus*-FFRFamide from *Mytilus edulis* [6] and the *Helix* crustacean cardioactive peptide (CCAP)-related peptide-III from *Helix pomatia* [7]. They all contain D-amino acids at position 2. Like other molluscs, these species have been extensively studied with regard to neurophysiology, immunocytochemistry, neurobiology *etc.* and many kinds of neuropeptides consisting only of L-amino acids have been isolated as well. On the other hand, contryphan from *Conus radiatus* has a characteristic feature [8]. The octapeptide, which has a D-tryptophan at position 4, has been isolated from the venom of the fish-hunting snail and acts exogenously with other toxins to cause an excitotoxic shock response in the prey. This chapter will therefore focus on the purification, structural determination, molecular conformation, biological activities, precursor protein and localization of molluscan neuropeptides having D-amino acids at position 2.

Purification

All of the peptides isolated from mulluscs as neuropeptides containing D-amino acids consist of 4 to 12 amino acid residues. Purification of these rather small peptides is generally performed using a combination of different high-pressure liquid chromatography (HPLC) systems such as gel-permeation, ion-exchange or reversed-phase HPLC (RP-HPLC). Moreover, their chromatographic behavior is an aid in identification and structural determination. At least one of the retention times of such peptides on several HPLC systems is commonly different from that of their L-isomeric peptides or other D-isomeric peptides. In this section, purification procedures, specifically concentrating on the neuropeptides from *A. fulica*, will be described.

Achatin-I

A. fulica belongs to the gastropods of the molluscs, originating in East Africa and now found abundantly in South-East Asia. The land snails used were collected in Manila, Philippines or in Okinawa, Japan. Achatin-I (Gly-D-Phe-Ala-Asp) was originally purified from the cerebral and suboesophageal ganglia of *Achatina* using the bioassay system on the periodically oscillating neuron (PON) of the snail [1, 2]. The same peptide has since been isolated from *Achatina* atria using the atrium and the ventricle of the snail as a bioassay system [3]. Figure 1 shows the procedure for purifying achatin-I from the ganglia of the snail. The yield of achatin-I was about 50 μg, starting from 30,724 ganglia. During the course of purifying achatin-I from the ganglia, the L isomer of achatin-I, namely achatin-II (Gly-Phe-Ala-Asp) was also identified (about 17 μg from 30,724 ganglia). Achatins-I and -II were eluted at different retention times using these purification systems (Fig. 1). The peptide from the atria of *Achatina* was formerly named *Achatina* cardioexcitatory substance-1 (ACES-1) and was later found to be identical to achatin-I from the ganglia [3]. It has never been determined whether achatin-II is also located in the atria.

Fulicin

Purification of fulicin (Phe-D-Asn-Glu-Phe-Val-NH_2) was originally performed from the cerebral and suboesophageal ganglia of *Achatina* [4]. The scheme for the purification of fulicin from the central ganglia of 1000 snails is shown in Figure 2. The peptide was isolated from the ganglia using the bioassay system on the penis retractor muscle of the snail. An all-L-isomeric fulicin has been never found in the ganglia.

Recently, both fulicin and its L isomer were identified in the extract of atria using the synthetic peptide as a marker [5]. This procedure was the

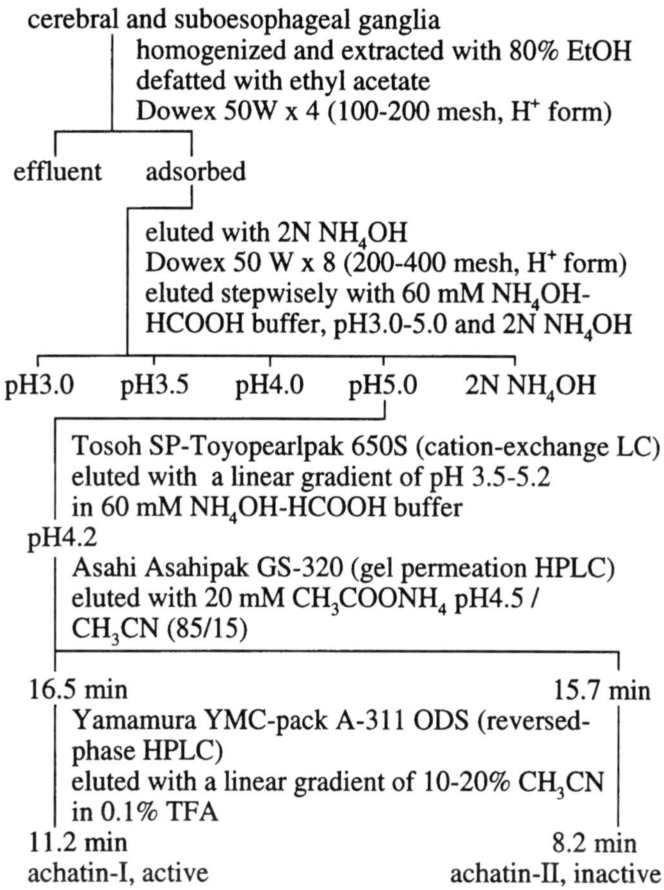

Figure 1. Schematic procedure for the purification of achatin-I, II from the cerebral and suboesophageal ganglia of *A. fulica*. (From refs 1, 2).

same as that used for the purification of fulyal described below (Fig. 3). The amounts of fulicin and L-isomeric fulicin purified from 5 g of frozen atria were about 120 pmol and 780 pmol, respectively.

Fulyal

The sequences of other fulicin-gene related peptides (FGRPs) have been predicted from the complementary DNA (cDNA) encoding the fulicin precursor protein in the ganglia of *Achatina* [9] as described in the section on precursor proteins. RNA blot analysis has indicated that single transcripts of the fulicin precursor exist in the ventricles and atria as well as in the ganglia [9]. In order to clarify whether FGRPs exist in the atria and whether they

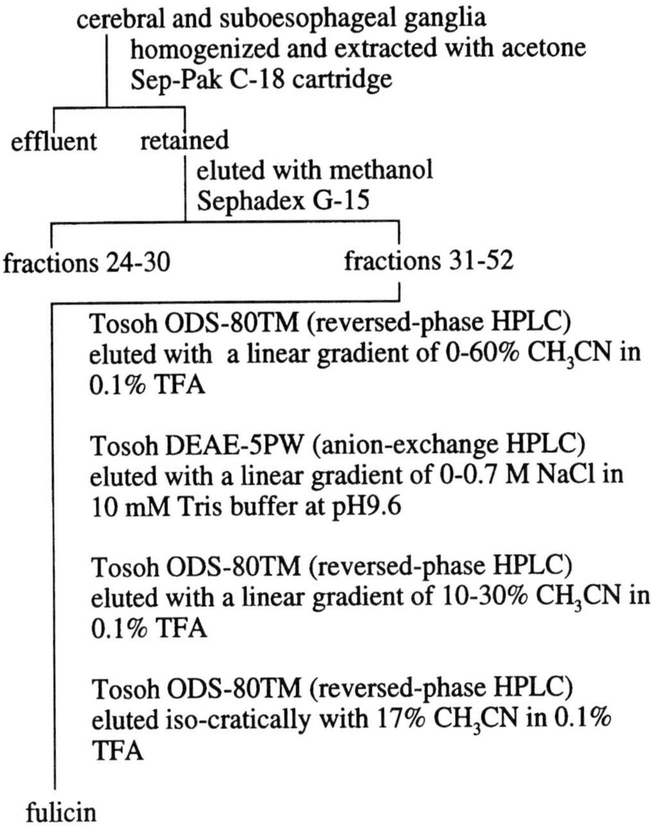

Figure 2. Schematic procedure for the purification of fulicin from the cerebral and suboesophageal ganglia of *A. fulica*. (From ref. 4).

contain D-amino acids, we attempted to purify the predicted peptides from the atria [5]. Previous to the extraction of the atria, two series of fulicin and FGRP-1–9 were synthesized: (i) consisting of only L-amino acids; and (ii) containing a D-amino acid at position 2. Synthetic peptides were then characterized based on the retention time in RP-HPLC. Under the conditions used, all D-isomeric FGRPs were eluted at positions different from their L isomers except for fulicin, which was coeluted with its L isomer. The identification of natural peptides from the atria was achieved by comparing their chromatographic behavior with that of synthetic peptides. Figure 3 shows the procedure for purifying peptides from the atria. The fraction number in cation-exchange HPLC, retention time in RP-HPLC, structure and yield of the purified peptides from the atria are shown in Table 1. Among these purified peptides, only FGRP-9 (Tyr-Ala-Glu-Phe-Val-NH_2) was clearly shown to coexist with its D isomer (Tyr-D-Ala-Glu-Phe-Val-NH_2) in the *Achatina* atria like fulicin. This novel D-amino acid-containing

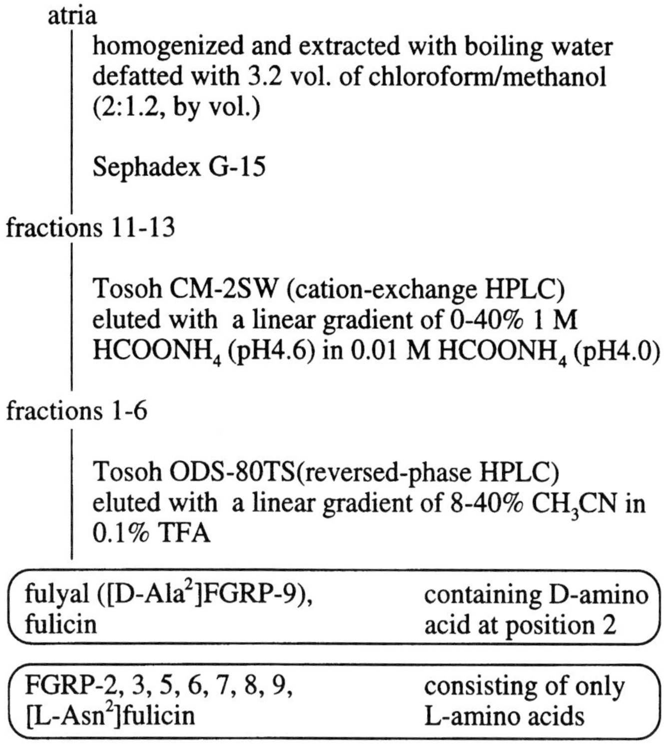

Figure 3. Schematic procedure for the purification of fulyal, fulicin and FGRPs from the atria of *A. fulica*. (From ref. 5).

peptide was named "fulyal": "ful" was derived from fulicin, and "yal" is the one-letter notation of the amino acid residues, Tyr, Ala and Leu, that differ from those of fulicin. The yields of fulyal and its L isomer from *Achatina* atria (5 g) were about 140 pmol and 90 pmol, respectively.

Mytilus-FFRFamide

The neuropeotide FMRFamide (Phe-Met-Arg-Phe-NH_2) was originally isolated from the central nervous system of the clam *Macrocallista nimbosa* [10]. The peptide and various related peptides are widely distributed in a number of species belonging to different classes of molluscs. *Mytilus edulis* belongs to the bivalve molluscs. The anterior byssus retractor muscle (ABRM) of the marine mussels has been found to contain at least two species of FMRFamide-related peptides, namely FMRFamide itself and *Mytilus*-FFRFamide (Ala-D-Leu-Ala-Gly-Asp-His-Phe-Phe-Arg-Phe-NH_2) [6]. These peptides were isolated from 581 g of *Mytilus* ABRM using the bioassay system [6]. The L isomer of *Mytilus*-FFRFamide has not yet been isolated.

Table 1. Fulyal, Fulicin and FGRPs from *Achatina* atria

Fraction no.	Peak no.	Retention time (min)	Proposed structure	Identification	Purified amount (pmol)
1	a	20.6	Ala-Glu-Phe-Leu-OH	des-Tyr[1]-FGRP-9-OH	430
	b	21.4	Phe-[Asn]-Glu-Phe-Val-NH_2	fulicin	120
	c	22.3	Ser-Pro-Tyr-Asp-Phe-Val-NH_2	FGRP-8	320
2	d	21.1	Asn-Tyr-Asp-Phe-Val-NH_2	FGRP-7	ND
			Ser-Tyr-Asp-Phe-Val-NH_2	FGRP-2	mixture
	e	21.4	Phe-Asn-Glu-Phe-Val-NH_2	[L-Asn[2]]fulicin	100
	f	26.6	Tyr-Ala-Glu-Phe-Leu-OH	FGRP-9-OH	60
3	g	21.4	Phe-Asn-Glu-Phe-Val-NH_2	[L-Asn[2]]fulicin	420
	h	24.8	Ser-Pro-Tyr-Asp-Phe-Ile-NH_2	FGRP-6	590
4	i	21.4	Phe-Asn-Glu-Phe-Val-NH_2	[L-Asn[2]]fulicin	260
	j	23.5	Tyr-Asp-Phe-Ile-NH_2	FGRP-5	ND
	k	24.2	Tyr-[Ala]-Glu-Phe-Leu-NH_2	[D-Ala[2]]FGRP-9 (fulyal)	140
5	l	23.5	Tyr-Ala-Glu-Phe-Leu-NH_2	FGRP-9	50
	m	25.2	Thr-Tyr-Asp-Phe-Leu-NH_2	FGRP-3	70
6	n	23.5	Tyr-Ala-Glu-Phe-Leu-NH_2	FGRP-9	40

Fraction no., fraction number of the cation exchange HPLC; peak no., peak number in the RP-HPLC; retention time, retention time of the RP-HPLC; purified amount, estimated amount of each purified peptide based on amino acid analysis; boxed letters, D-amino acid. (From ref. 5).

Helix *CCAP-related peptide-III (HCCAP-RP-III)*

The crustacean cardioactive peptide (CCAP) was originally purified from the pericardial organs of *Carcinus* as one of several cardioactive substances [11]. Minakata et al. isolated three peptides from the ganglia of the European land snail, *Helix pomatia* [7]. These peptides resembled CCAP and were named *Helix* CCAP-related peptides (HCCAP-RPs)-I, II and III. Among the purified peptides. HCCAP-RP-III (Leu-D-Phe-Cys-Asn-Gly-Tyr-Gly-Gly-Cys-Gln-Asn-Leu-NH_2, disulfide bond between Cys^3 and Cys^9) was found to contain a D-Phe at position 2 and to be a D-isomeric peptide of HCCAP-RP-II.

Structural determination

D, L-amino acid analysis – using a liquid chromatographic method with chiral reagents such as *o*-phthaldialdehyde (OPA) together with *N*-isobutyryl-L-cysteine (IBLC) [12] or (+)-and (–)-1-(9-fluorenyl)ethyl chloroformate (FLEC) [13] – is becoming one of the routine analyses in the structural determination of the peptides containing D-amino acids. In this section, the latest method employed for fulyal from *Achatina* atria [5] will be briefly mentioned.

As described above, the sequences of FGRPs have been predicted from the cDNA encoding the fulicin precursor protien [9]. Comparing the retention times of the synthetic peptides, all peaks that emerged during the 19–30 min of the last step of the purification system using RP-HPLC (Fig. 3), were collected and sequenced. Fourteen peaks (*a–n*) were then recognized to contain peptides related to FGRPs (Tab. 1). Sequencing of peaks *k*, *l* and *n* showed a common sequence of Tyr-Ala-Glu-Phe-Leu (FGRP-9). Their C-termini were identified as the α-amidated form based on mass spectrometry [observed $(M + H)^+ = 641.3-641.5$]. To confirm D-and L-amino acid compostion, the peptides were subjected to amino acid analysis using OPA together with IBLC [12] (Fig. 4). Among them, the Ala residue in peptide *k* was determined to be the D-amino acid; thus peptide *k* was [D-Ala^2]FGRP-9 (fulyal). Peptides *l* and *n* consisted entirely of L-amino acid residues and proved to be L-isomeric FGRP-9. The structures of other peptides from *Achatina* atria were confirmed in the same manner (Tab. 1).

Molecular conformation

Among the molluscan neuropeptides that contain D-amino acids, achatin-I is the only one whose three-dimensional structure is based on its X-ray crystal analysis [14–16]. Analysis has indicated that the achatin-I molecule

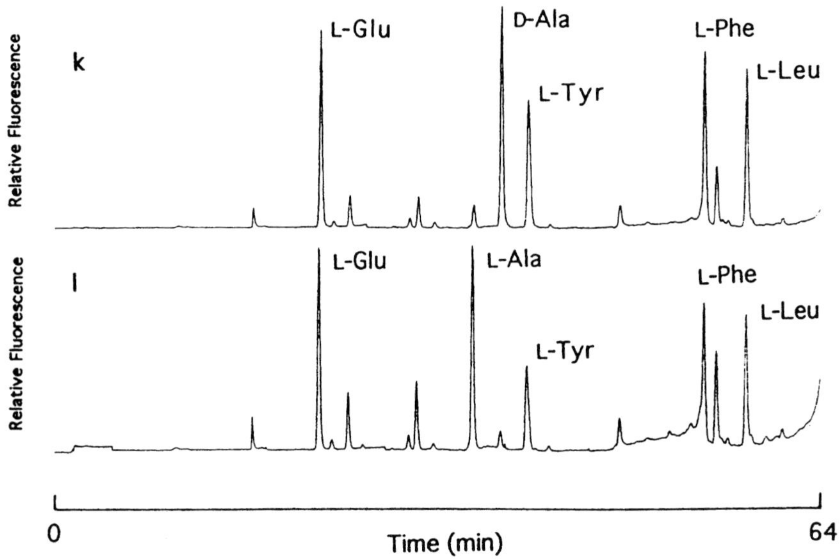

Figure 4. HPLC profile of amino acids generated after hydrolysis and derivatization with OPA and IBLC of purified materials (*k* and *l* in Tab. 1) derived from *Achatina* atria. Separation, RP-HPLC; column, CAPCELL PAK C18 UG 120 (Shiseido, Ltd., 4.6 mm × 25 cm, 5 μm); mobil phases, 5% (*A*) and 60% (*B*) of methanol/acetonitrile (6:4, v/v) in 0.02 M phosphate buffer pH 5.75; gradient conditions, 0–75% B in A over 1 h; flow rate, 1 ml/min; fluorescence detection, excitation at 340 nm and emission at 450 nm. (Reprinted with permission from ref. 5).

exists as a zwitterionic state with a protonated N-terminal end and a deprotonated β-carboxyl group on the C-terminal Asp side chain. As shown in Figure 5, the molecule takes a turn conformation stabilized by the two intramolecular hydrogen bonds of NH(Gly)...C^{γ}=O^{δ}(Asp) and NH(Asp)...C^{γ}=O^{δ}(Asp), thus forming a 15-membered ring structure. On the other hand, achatin-II has an extended conformation (Fig. 5). Achatin-II appears also to exist in zwitterionic form; however, the chemical state of achatin-II is different from that of achatin-I. In the achatin-II molecule, the N-terminal end was protonated, and the α-carboxyl group of the C-terminal end was deprotonated, not the β-carboxyl group. The extended conformation of achatin-II is in contrast with the turn conformation of achatin-I. The replacement of D-Phe with L-Phe (achatin-I → II) would cause severe steric hindrance between the Phe benzene ring and the backbone chain, leading to the extended conformation. The conformational comparison indicates the structural role which the D-Phe residue of achatin-I plays in constructing a definite active form as described below.

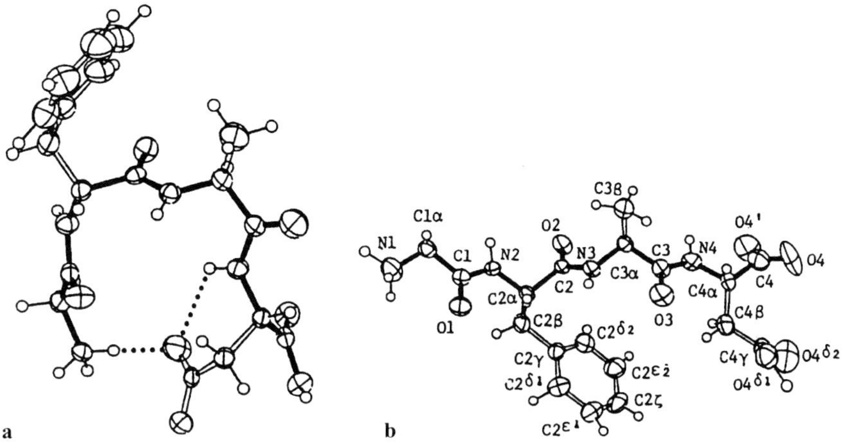

Figure 5. Conformational comparison of achatin-I (a) and achatin-II (b) molecules observed in their crystal structures. The dotted lines in (a) represent intramolecular hydrogen bonds. (Reprinted with permission from ref. 16).

Biological activity

In general, molluscan neuropeptides, such as the FMRFamide family, affect several aspects of the electrophysiological characteristics of neurons and muscles [17]. These peptides act as transmitters and neuromodulators and also as neurohormones. Peptides having the D-amino acids described here are apparently no exception, and play various important physiological roles as neuropeptides.

Effects on neurons

Achatin-I is the only molluscan D-amino acid-containing neuropeptide whose physiological and pharmacological features in neurons have been investigated in detail. Figure 6 shows the effect of 10^{-5} M achatin-I on PON, which is the gaint neuron in the suboesophageal ganglion of *Achatina*, used in the bioassay system [1, 2]. Achatin-I produces marked depolarization and an inward current in the neuron. The analysis of the dose-response curve reveals that the ED_{50} is about 2.0×10^{-6} M. The effect of achatin-I is blocked in the Na^+-free state, indicating Na^+ dependency [1, 2]. Achatin-I shows marked excitatory effects not only on many other neurons in the suboesophageal ganglia, such as TAN (tonically autoactive neuron and VIN (visceral intermittently firing neuron), but also on neurons in the cerebral ganglia, v-RCDN (ventral-right cerebral distinct neuron) and -RCDN (dorsal-right cerebral distinct neuron) [18]. Among the 23 types of neurons tested, 10 types are excited by achatin-I, whereas no neurons is

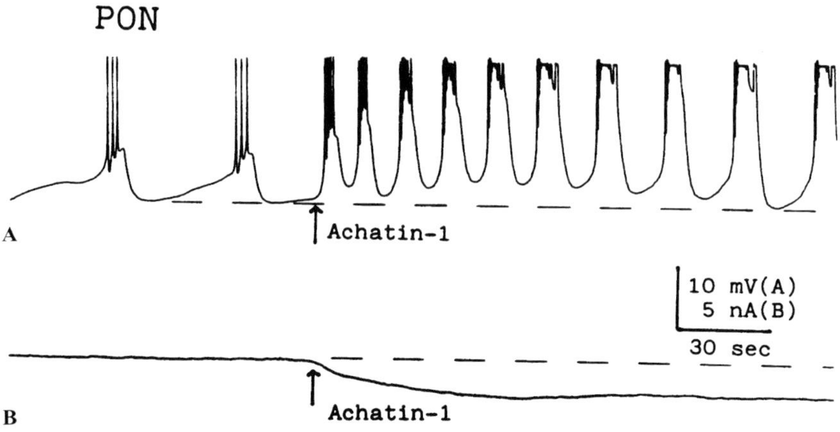

Figure 6. Effect of achatin-I on the giant neuron PON in the suboesophageal ganglion of *A. fulica*. The peptide was applied by bath method at 10^{-5} M (arrow). (A) Current clamp; (B) voltage clamp. Vertical bar, calibration (10 mV for (A) and 5 nA for (B); horizontal bar, time course (30 s). (Reprinted with permission from ref. 2).

inhibited [18]. Thus achatin-I has been proposed as an excitatory neurotransmitter of *Achatina* neurons. On the contrary, the L-isomeric coexisting peptide, achatin-II, has been found to be entirely inactive in all neurons tested. Among the other synthetic diastereomers containing D- or L-amino acid residues, only [D-Ala[3]]achatin-I had slight excitatory effects [2, 18]. Various synthetic analogs of achatin-I have also been examined for their effects. However, only Gly-Gly-D-Phe-Ala-Asp, D-Phe-Ala-Asp and Gly-D-Phe-Ala-Asn showed excitatory effects, and their potencies were much less compared with achatin-I [19]. These results have indicated that the sensitivity of neurons to achatin-I is strictly structure-specific as well as stereospecific [20]. In addition, achatin-I is thought to act as a neuromodulator because it shows both suppressing and enhancing actions on the excitatory or inhibitory effects of transmitters, such as oxytocin, APGWamide (Ala-Pro-Gly-Trp-NH_2), FMRFamide or acetylcholine, in *Achatina* neurons [20].

Fulicin also appears to modulate the activity of the neurons of *Achatina* [4]. The peptide produces rhythmic bursts of firing in neurons B1 and B4, which are located in the buccal ganglion and control the contractions of the buccal muscles.

Effects on heart

Although achatin-I has been isolated from the atria of *Achatina*, it has no direct prominent effects on the atrium; instead, the peptide shows excitatory effects on the ventricle (Fig. 7) [3]. Furukawa and Kobayashi [21, 22]

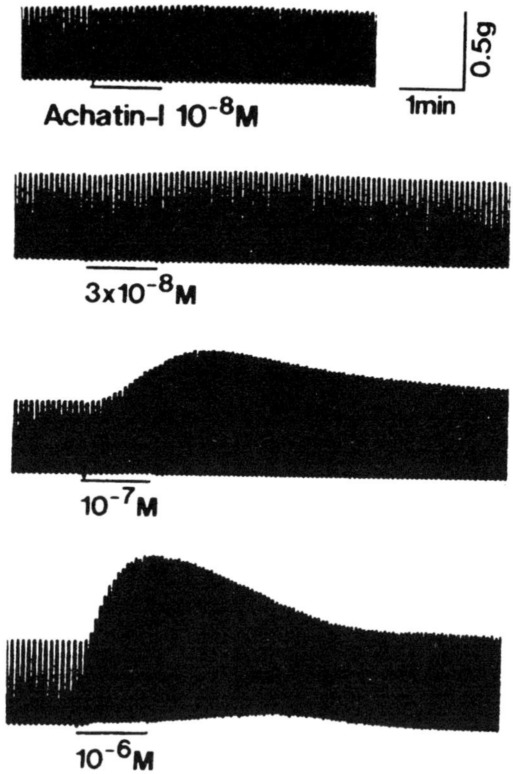

Figure 7. Dose-dependent excitatory effects of achatin-I on the ventricle of *A. fulica*. The peptide was applied during the period shown by the horizontal line under each record. (Reprinted with permission from ref. 3).

have extensively investigated the neural control of the heart beat in *Achatina*, and seven heart regulatory neurons [PON, TAN, TAN-2, TAN-3, d-RCDN, d-LCDN (dorsal-left cerebral distinct neuron) and VG1 (visceral ganglion neuron)] have been identified in the central nervous system. Among these neurons, PON is the most effective heart excitator, producing heart excitation at rather low firing frequencies. As described above, achatin-I expresses excitatory influences not only on ventricles but also on heart excitatory neurons, i.e., PON, TAN and d-RCDN, in the central ganglia. Moreover, the effects of achatin-I on a preparation of central ganglia and atrium connected by the intestinal nerve has been examined [23]. Application of achatin-I to the ganglia causes depolarization and superimposed spikes in PON accompanied by prominent enhancement of atrial beat. These observations suggest that achatin-I is involved in the regulation of atrial and ventricular beat by centrally or peripherally acting as an excitatory neuromodulator.

As shown in Figure 7, achatin-I enhances both the frequency and amplitude of the ventricular beat with the threshold at about 10^{-8} M, while the seven other synthetic combinations of L and D isomers of achatin-I show no significant effects on the ventricle up to 10^{-5} M [3]. The effect of the all-L-isomeric peptide, achatin-II, is about 1×10^{-4} that of achatin-I. Moreover, a number of analogs of achatin-I have been synthesized, and the structure-activity relations have been investigated on the ventricle [23]. Replacement of the D-Phe residue at position 2 with D-Tyr or D-Leu reduces the activity to 1×10^{-2} to 1×10^{-3} that of achatin-I. These observations indicate that D-Phe^2 is essential for the activity.

Similarly, fulicin shows modulatory action on the ventricular muscles of *Achatina*, though it has been isolated from the atria, like achatin-I [4]. At a concentration of 10^{-6} M fulicin enhances the amplitude of the beat, but with no significant change in the freqency, unlike achatin-I.

Effects on muscles

There is much evidence that a main transmitter and one or more neuromodulators are coreleased to modulate transmission in the molluscan neuromuscular system. The penis retractor muscle of *Achatina* also appears to be regulated by several neuropeptides [3–5, 24–26]. Kobayashi et al. [26] have examined the effects of putative neurotransmitters and modulators on the muscle. Among the substances tested, the neuropeptide fulicin was the most effective in enhancement of the tetanic contractions evoked by electrical stimulations; the threshold was extremely low (10^{-12} M). Recently, the activity of the neuropeptide fulyal was examined and shown to be as high as that of fulicin (threshold below 10^{-12} M) (Fig. 8) [5]. The other D-amino acid-containing neuropeptide from *Achatina*, achatin-I, also enhances the tetanic contraction; however, the threshold is 10^{-9} M or greater [3]. As for the neuropeptides consisting only of L-amino acids, FGRP-7 (Asn-Tyr-Asp-Phe-Val-NH_2) is the most active peptide [5]. Its threshold is 10-fold higher than that of fulyal or fulicin (10^{-11} M); however, other peptides such as ACEP-1 [24], WWamides [27] and AEP-2,3 [28] are at least 100-fold less active than FGRP-7. Among the so-called classical transmitters, 5-hydroxytryptamine is the most effective potentiator of evoked contractions at concentrations of 10^{-9} M or higher [26]. At present, therefore, neuropeptides containing the D-amino acids fulicin and fulyal appear to be the most potent peptides in enhancing contraction of the penis retractor muscle of *Achatina*. In addition to the penis retractor muscle, the effects by achatin-I and fulicin on the radula retractor muscle of *Achatina* have been examined. Achatin-I increases the amplitude of the tetanic contraction in response to electrical stimulations, with a threshold of 10^{-8} M [3]. In contrast, fulicin at 10^{-8} M or more depresses tetanic contraction [4].

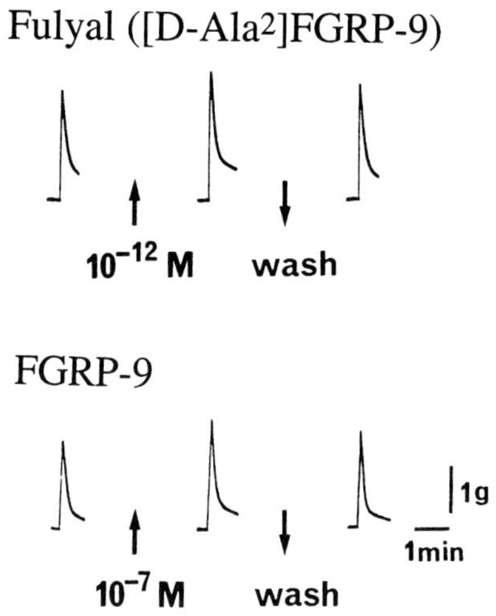

Figure 8. Effects of fulyal ([D-Ala^2]FGRP-9) and FGRP-9 (all-L-isomer of fulyal) on the penis retractor muscle of *A. fulica*. Tetanic contraction of the muscle was evoked by repetitive electrical stimulations (20 V, 0.6 ms, 40 Hz, for 1 s). The peptide was applied to the muscle 8 min prior to the stimulation (upward arrow) and washed out soon thereafter (downward arrow). (Reprinted with permission from ref. 5).

The structure-activity relationships of fulicin have also been examined on the contraction of the penis retractor muscle of *Achatina* [29, 30]. The all-L-isomer is 3000-fold less active than fulicin, whereas substitution of Gly or the other D-amino acid for D-Asn^2 shows no significant reduction in activity. In contrast, D-isomeric peptides having one D-amino acid each at positions 1, 3, 4, or 5 are at least 20,000-fold less active than fulicin. D-chirality at position 2 appears to be very important for activity [30]. As for fulyal, such detailed investigations have not yet been performed; however, fulyal has a structure analogous to fulicin. As described below, fulicin and fulyal ([D-Ala^2]FGRP-9) have a similar feature in the N-terminus (aromatic-small and neutral-Glu-) as well as in the C-terminus, which is common to fulicin and all FGRPs (-Phe-Val/Ile/Leu-NH_2). Moreover, the threshold concentration of the L-isomeric fulyal is about four orders of magnitude higher than that of fulyal (Fig. 8). The influence of D-amino acid on the activity is common to fulyal and fulicin. Although the molecular conformations of fulyal and fulicin remain unknown, they may differ from those of their L-counterparts. As in the case of achatin-I, the conversion from the L- to D-amino acid residue in such short peptides affects the entire molecular structure. The change in the molecular structure would have a significant influence on biological activity.

However, reports on other molluscan peptides containing D-amino acids indicate that there are at least two other cases for the impact of D-amino acids on biological activity in sequences. One is the case that conversion from the L- to the D-amino acid residue has no influence on activity and the other is that conversion causes a decrease in activity. The former can be seen in *Mytilus*-FFRFamide [6] and the latter in HCCAP-RP-III [7]. *Mytilus*-FFRFamide containing D-Leu^2 shows potentiation phasic contraction of ABRM elicited by repetitive electrical stimulation [6]. Several isomers and peptide fragments have been synthesized and their effects examined. The all-L-isomeric peptide as well as [D-Ala^1, L-Leu^2]*Mytilus*-FFRFamide show effects on the muscle almost equipotent to the native peptide. It appears that the C-terminal tetrapeptide is very important to FMRFamide-like excitatory action on the ABRM and that the N-terminal hexapeptide portion is not essential to activity. As for HCCAP-RP-III [7], the all-L-isomeric peptide, namely HCCAP-RP-II, shows potent myotropic activity on phasic contractions of the ABRM of *Mytilus* (threshold concentration, 10^{-11} M). In contrast, HCCAP-RP-III containing D-Phe^2 is about 10,000-fold less active than HCCAP-RP-II. In the radula retractor and protractor muscles of *Rapana thomatiana*, HCCAP-RP-II shows significant excitatory effects on twitch contractions evoked by electrical stimulation; however, HCCAP-RP-III shows no activity in either muscle at concentrations as high as 10^{-5} M. No observable activity has yet been found on HCCAP-RP-III. The stereomeric effects on HCCAP-RP-III are contrary to those on achatin-I, fulyal and fulicin.

In general, the occurrence of D-amino acids in peptides is considered to increase their resistance to proteinase. Additionally, D-amino acids in achatin-I, fulicin and fulyal may help to shape the specific conformation for each biological activity. However, in the case of *Mytilus*-FFRFamide or HCCAP-RP-III, the contribution of the D-amino acid is possibly very low, or the D-amino acid may inhibit formation of a suitable structure for exerting the activity. These observations led to the hypothesis that, at first, small peptides containing D-amino acids appeared as a mechanism for resisting proteinase, then, that their receptors changed to recognize their specific conformations. This seems to be a reasonable assumption at least in the case of *Achatina* D-amino acid-containing neuropeptides. In order to discuss this hypothesis, it will be necessary to find more novel peptides having D-amino acids, to examine their biological activity and to elucidate the nature of their corresponding receptors.

Precursor proteins

Until recently, it was unknown whether molluscan D-amino acid-containing neuropeptides are synthesized through a ribosomal pathway or through a nonribosomal pathway [9]. However, it seemed possible that molluscan neuropeptides are synthesized by the same mechanism as that of amphibian opioid peptides, because all of the molluscan peptides have D-amino acid residues at position 2, as in the case of amphibian peptides [9, 31, 32]. To verify the above posibility, an attempt was made to clone the cDNA that encodes the precursor proteins of achatin-I and fulicin. A novel cDNA encoding the fulicin precursor protein was cloned from snail cerebral and suboesophageal ganglia with a synthetic DNA probe corresponding to the fulicin amino acid sequence followed by Gly (FNEFVG) [9] (GenBank/EMBL/DDBJ Data Bank, accession no. D13986). The amino acid sequence of the precursor deduced suggests that fulicin is synthesized through a ribosomal pathway and D-Asn is coded by the usual L-Asn codon (AAT). As shown in Figure 9, the predicted fulicin precursor protein (357 amino acids) contains one copy of fulicin and at least nine other putative α-amidated neuropeptides composed of four to six amino acid residues. These putative neuropeptides are called fulicin gene-related peptides (FGRPs)-1 to -9 of which seven peptides are novel and two peptides are the same structures as *Mytilus* inhibitory peptide (MIP)-related peptides (MIPRPs) previously isolated from *Helix pomatia* [9, 33]. Although fulicin and FGRPs have a similar feature in the C-terminal residues (Phe-Val/Ile/Leu-NH_2), they can be classified into three groups. The first group is FGRP-4, 6 and 8 which can be regarded as MIPRPs. A significant feature of the MIP family is a Pro residue that is considered to be very important for its inhibitory action on the ABRM [33–35]. FGRP-1, 2, 3, 5 and 7

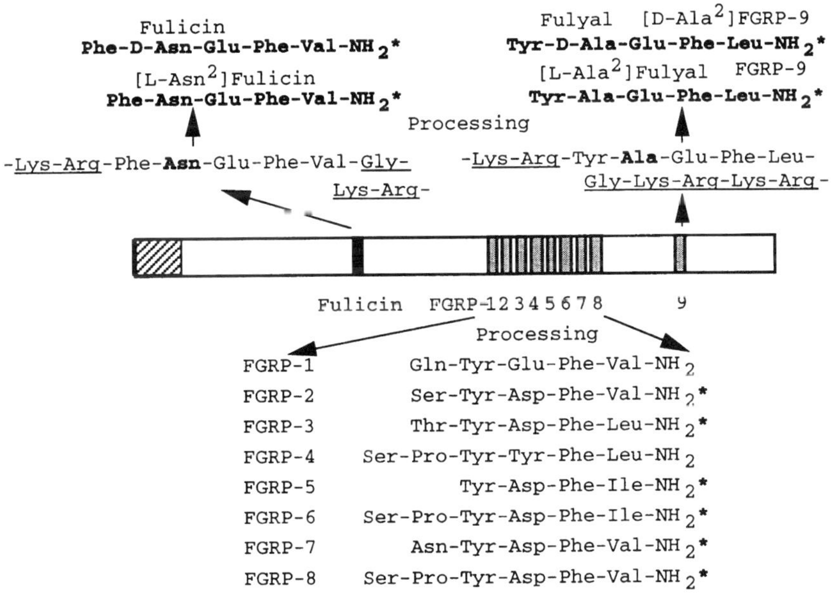

Figure 9. Schematic representation of the fulicin precursor protein and its processing products from *A. fulica*. The hatched box represents the putative N-terminal signal peptide. The fulicin (black box) and FGRP (stippled boxes) sequences are flanked by Lys-Arg (Lys) at the N-terminus and by Gly-Lys-Arg (Lys) at the C-terminus. The asterisks represent the peptides which have been purified from the atria of *A. fulica*. (From ref. 5).

make up the second group. They have an acidic amino acid in the third position and Tyr in the fourth position from the C-terminus. The members of the third group are fulicin and FGRP-9, because the fourth position is not Tyr and their N-terminal amino acids are aromatic.

Although FGRPs-1 to -8 are consecutively located, fulicin and FGRP-9 are located apart from this region (Fig. 9). All sequences of the 10 peptides are flanked by Lys-Arg at the N-terminus and by Gly-Lys-Arg at the C-terminus except for one position between FGRP-6 and FGRP-7 that is flanked by Gly-Lys-Lys. The Gly-Lys-Arg/Lys sequence is a consensus signal for processing and subsequent amidation in the precursor proteins of various bioactive peptides. Processing endoprotease or prohormone convertase (PC), involved in the processing of such proproteins, has never yet been elucidated in *Achatina*. However, using the polymerase chain reaction (PCR), a cDNA product (141 base pairs) that encodes a sequence which is the most homologous to neuroendocrine convertase of PC2 has been cloned from *Achatina* ganglia. The PCR product (GenBank/EMBL/DDBJ Data Bank, accession no. D17459) has 78.7% identity with PC2 from human, mouse, pig and rat. The putative PC, of which a fragment is encoded by the PCR product, is a candidate processing endoprotease which is involved in the processing of the fulicin precursor protein. Besides, as described above, a novel D-amino acid-containing peptide, named fulyal ([D-Ala^2]FGRP-9), has been isolated from the atria of *Achatina* together with its L isomer as well as fulicin, the L isomer of fulicin, FGRP-2, 3, 5, 6, 7 and 8 [5]. These findings indicate that fulicin, fulyal and related peptides are produced in the snail by processing of the ribosomally made precursor described here, and that L-isomeric fulicin and fulyal further undergo epimerization to yield D-isomeric fulicin and fulyal. Although the mechanism of the in-chain epimerization remains unclear, the following questions merit consideration: Why does the snail produce such closely related D-amino acid-containing peptides bearing the same muscle contraction-enhancing activity? And why are fulicin and FGRP-9 located apart from the region where FGRP-1 to -8 are consecutively grouped? In addition, the sequence of the achatin-I precursor protein is still unknown. However, achatin-I seems also to be flanked by the dibasic sequence Lys-Arg in large precursor proteins like fulicin and fulyal, since achatin-I and II together with fulicin and fulyal are present both in the ganglia and atria of *Achatina*. If this is the case, it is possible to clone the cDNA that encodes the achatin-I precursor protein with a synthetic DNA probe corresponding to the achatin-I amino acid sequence linked to Lys-Arg.

Localization

Studies on the localization and distribution of neuropeptides have become an important tool in neurobiology, since these studies, in conjunction with biological, physiological and pharmacological investigations, have the potential to elucidate the physiological role of peptides and the molecular mechanisms that underly the cell-to-cell communication network. Such studies have been performed only on fulicin and fulyal among the molluscan peptides having D-amino acids.

To examine the localization of fulicin, an immunohistochemical technique was employed [26, 29]. In the cerebral ganglia of *Achatina*, a number of immunoreactive neurons were observed in the right and left caudal quadrants of the right and left ganglia, respectively. These neurons occurred in several clusters consisting mostly of small neurons (about 10 μm in diameter) and a few larger ones (about 50 μm). In the suboesophageal ganglia, several neurons, mainly in the right pedal and right pleural ganglia, were immunoreactive. Fulicin-like immunoreactivities were also found in the cerebropleural, cerebropedal connectives, numerous nerve fibers and the penis retractor muscle.

Northern blotting indicated that single transcripts of the fulicin precursor existed in the suboesophageal and cerebral ganglia as well as the ventricles and the atria [9]. Fulicin and possibly fulyal are located in the central ganglia and atria, and exhibit modulatory actions on the activity of the penis retractor and ventricular muscles, as well as several other muscles and neurons. These investigations suggest that some fulicin, fulyal and related peptides might be produced in the neuron and some of them might be synthesized in the heart itself. Moreover, localization of fulicin messenger RNA (mRNA) in the *Achatina* ganglia has been investigated by *in situ* hybridization [36]. There were several neurons expressing the messenger encoding the fulicin precursor protein in both the suboesophageal and cerebral ganglia. In the cerebral ganglia, positive cells consisted of a few larger neurons (30–40 μm in diameter) and a cluster of several smaller ones (< 10 μm) (Fig. 10). In the suboesophageal ganglia, several positive cells (20–40 μm) existed in the right parietal ganglion. These results provide the first evidence that fulicin and fulyal are indeed biosynthesized in specific nerve cells in both the cerebral and suboesophageal ganglia. Positive cell *in situ* hybridization analysis reflects the exclusive site of fulicin precursor biosynthesis. On the other hand, the intensity of fulicin-like immunoreactivity depends on the amount of fulicin-like peptides released and transported from the secretory cells. It is therefore reasonable to assume that immunoreactivity is not necessarily in agreement with the expression of the gene. Still, positive neuron *in situ* hybridization analysis seemed to be less than expected from the immunohistochemical experiments. This difference might be caused in part by the rather broad specificity of the antibody used in the immunohistochemical study.

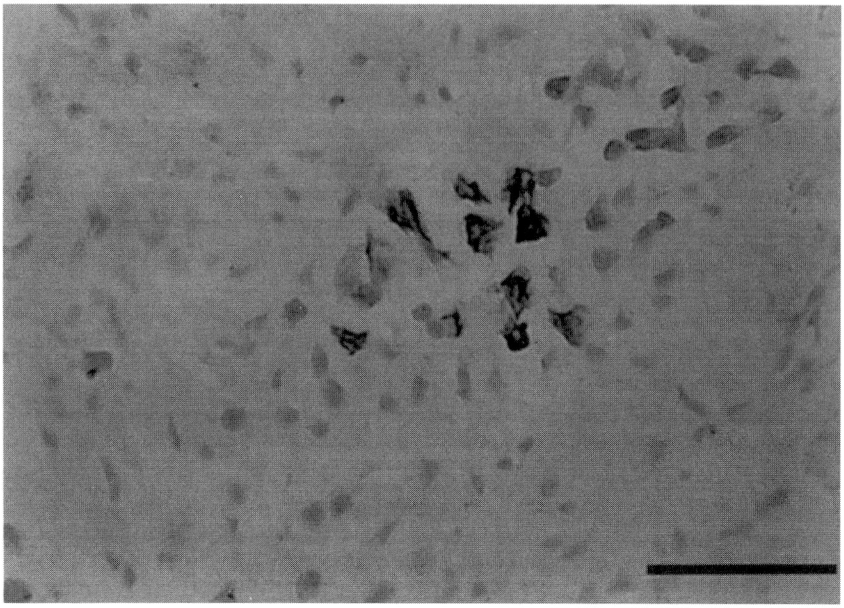

Figure 10. Localization of fulicin mRNA in the cerebral ganglia of *A. fulica* by *in situ* hybridization. Caribration bar, 50 μm. There was a cluster of several positive smaller neurons (10 μm) in the center of this figure. cRNA probe was generated from the clone encoding the fulicin precursor cDNA, labelled with digoxigenin-UTP and detected using an antidigoxigenin horseradish peroxidase-conjugated antibody (Boehringer Mannheim Biochemicals). The color was developed with diaminobenzidine tetrahydrochloride and counterstained with methylgreen (From ref. 36).

Conclusions

Many of the observations presented here suggest that snails and bivalves in molluscs make dynamic use of D-amino acids for neuropeptide synthesis. These neuropeptides exert excitatory or inhibitory effects on muscles and neurons, and serve as messengers both centrally and peripherally. However, it seems that they have a wider range of action. There is still much to be established about the physiological role of these interesting peptides. Further physiological, pharmacological and molecular biological studies on neuropeptides will help to determine the definitive role of these peptides, and investigation of their corresponding receptors will help us to understand the significance of the D-amino acids. In owing to the presence of D-amino acids in peptide chains and the molecular evolution of these peptides, it is expected that the neuropeptides achatin-I, fulicin, fulyal, *Mytilus*-FFRFamide and HCCAP-RP-III, and/or other novel peptides having D-amino acids will be discovered in other living organisms as well.

Acknowledgments
I would like to thank Drs. Makoto Kobayashi, Yojiro Muneoka, Masahiko Nakamura, Fumio Sakiyama, Hiroyuki Minakata and Kyosuke Nomoto for the guidance they have given me over the years.

References

1. Kamatani, Y., Minakata, H., Kenny, P.T.M., Iwashita, T., Watanabe, K., Funase, K., Ping Sun, X., Xongsiri, A., Kim, K.H. and Novales-Pi, P. et al. (1989) Achatin-I, an endogenous neuroexcitatory tetrapeptide from *Achatina fulica* Férussac containing a D-amino acid residue. *Biochem Biophys Res Commun* 160: 1015–1020.
2. Kamatani, Y., Minakata, H., Nomoto, K., Kim, K.H., Yongsiri, A. and Takeuchi, H. (1991) Isolation of achatin-I, a neuroactive tetrapeptide having a D-phenylalanine residue, from *Achatina* ganglia, and its effects on *Achatina* giant neurones. *Comp Biochem Physiol* 98C: 97–103.
3. Fujimoto, K., Kubota, I., Yasuda-Kamatani, Y., Minakata, H., Nomoto, K., Yoshida, M., Harada, A., Muneoka, Y. and Kobayashi, M. (1991) Purification of achatin-I from the atria of African giant snail, *Achatina fulica*, and its possible function. *Biochem Biophys Res Commun* 177: 847–853.
4. Ohta, N., Kubota, I., Takao, T., Shimonishi, Y., Yasuda-Kamatani, Y., Minakata, H., Nomoto, K., Muneoka, Y. and Kobayashi, M. (1991) Fulicin, a novel neuropeptide containing a D-amino acid residue isolated from the ganglia of *Achatina fulica*. *Biochem Biophys Res Commun* 178: 486–493.
5. Yasuda-Kamatani, Y., Kobayashi, M., Yasuda, A., Fujita, T., Minakata, H., Nomoto, K., Nakamura, M. and Sakiyama, F. (1977) A novel D-amino acid-containing peptide, fulyal, coexists with fulicin gene-related peptides in *Achatina* atria. *Peptides* 18: 347–354.
6. Fujisawa, Y., Ikeda, T., Nomoto, K., Yasuda-Kamatani, Y., Minakata, H., Kenny, P.T.M., Kubota, I. and Muneoka, Y. (1992) The FMRFamide-related decapeptide of *Mytilus* contains a D-amino acid residue. *Comp Biochem Physiol* 102C: 91–95.
7. Minakata, H., Ikeda, T., Muneoka, Y. and Nomoto, K. (1996) A new peptide containing a D-amino acid from the snail, *Helix pomatia*. Roles of D-amino acid residue. *Peptide Chem* 1995: 169–172.
8. Jimenéz, E.C., Olivera, B.M., Gray, W.R. and Cruz, L.J. (1996) Contryphan is a D-tryptophan-containing *Conus* peptide. *J Biol Chem* 271: 28002–28005.
9. Yasuda-Kamatani, Y., Nakamura, M., Minakata, H., Nomoto, K. and Sakiyama, F. (1995) A novel cDNA sequence encoding the precursor of the D-amino acid-containing neuropeptide fulicin and multiple α-amidated neuropeptides from *Achatina fulica*. *J Neurochem* 64: 2248–2255.
10. Price, D.A. and Greenberg, M.J. (1977) Structure of a molluscan cardioexcitatory neuropeptide. *Science* 197: 670–671.
11. Stangier, J., Hilbich, C., Beyreuther, K. and Keller, R. (1987) A novel cardioactive peptide (CCAP) from pericardial organs of the shore crab *Carcinus maenas*. *Proc Natl Acad Sci USA* 84: 575–579.
12. Brükner, H., Haasmann, S., Langer, M., Westhauser, T., Wittner, R., Godel, H. (1994) Liquid chromatographic determination of D- and L-amino acids by derivatization with *o*-phthaldialdehyde and chiral thiols. *J Chromatogr* 666: 259–273.
13. Einarsson, S. and Josefsson, B. (1987) Separation of amino acid enantiomers and chiral amines using precolumn derivatization with (+)-2-(9-fluorenyl)ethyl chloroformate and reversed-phase liquid chromatography. *Anal Chem* 59: 1191–1195.
14. Kamatani, Y., Minakata, H., Iwashita, T., Nomoto, K., In, K., Doi, M. and Ishida, T. (1990) Molecular conformation of achatin-I, an endogenous tetrapeptide containing D-amino acid residue. *FEBS Lett* 276: 95–97.
15. Ishida, T., In, Y., Doi, M., Inoue, M., Yasuda-Kamatani, Y., Minakata, H., Iwashita, T. and Nomoto, K. (1992) Crystal structure and molecular conformation of achatin-I (H-Gly-D-Phe-Ala-Asp-OH), an endogenous neuropeptide containing a D-amino acid residue. *Int J Peptide Protein Res* 39: 258–264.
16. Ishida, T., In, Y., Inoue, M., Yasuda-Kamatani, Y., Minakata, H., Iwashita, T. and Nomoto, K. (1992) Effect of the D-Phe^2 residue on molecular conformation of an endogenous neuropeptide achatin-I. *FEBS Lett* 307: 253–256.

17. Geraerts, W.P.M., Vreugdenhil, E. and Ebberink, R.H.M. (1988) Bioactive peptides in molluscs. In: M.C. Thorndyke and G.J. Goldsworthy (eds): *Neurohormones in invertebrates*. Cambridge University Press, New York, 261–281.
18. Kim, K.H., Takeuchi, H., Kamatani, Y., Minakata, H. and Nomoto, K. (1991) Slow inward current induced by achatin-I, an endogenous peptide with a D-Phe residue. *European J Pharmacol* 194:99–106.
19. Kim, K.H., Takeuchi, H., Kamatani, Y., Minakata, H. and Nomoto, K. (1991) Structure-activity relationship studies on the endogenous neuroactive tetrapeptide achatin-I on giant neurons of *Achatina fulica* Féussac. *Life Sci* 48:91–96.
20. Liu, G.J. and Takeuchi, H. (1993) Modulation of neuropeptide effects by achatin-I, an *Achatina* endogenous tetrapeptide. *Eur J Pharmacol* 240:139–145.
21. Furukawa, Y. and Kobayashi, M. (1987) Neural control of heart beat in the African giant snail, *Achatina fulica* Féussac. I. Identification of the heart regulatory neurones. *J Exp Biol* 129:279–293.
22. Furukawa, Y. and Kobayashi, M. (1987) Neural control of heart beat in the African giant snail, *Achatina fulica* Féussac. II. Interconnections among the heart regulatory neurones. *J Exp Biol* 129:295–307.
23. Kobayashi, M., Muneoka, Y., Fujimoto, K., Ohta, N., Yoshida, M., Kubota, I., Nomoto, K., Minakata, H. and Yasuda-Kamatani, Y. (1992) Structure and function of neuropeptides containing a D-amino acid residue isolated from the African giant snail *Achatina fulica*. In: A. Suzuki (ed.): *Peptide Chemistry*: Protein Research Foundation, Osaka, 321–326.
24. Fujimoto, K., Ohta, N., Yoshida, M., Kubota, I., Muneoka, Y. and Kobayashi, M. (1990) A novel cardio-excitatory peptide isolated from the atria of the African giant snail, *Achatina fulica*. *Biochem Biophys Res Commun* 167:777–783.
25. Harada, A., Yoshida, M., Minakata, H., Nomoto, K., Muneoka, Y. and Kobayashi, M. (1993) Structure and function of the molluscan myoactive tetradecapeptides. *Zool Sci* 10:257–265.
26. Kobayashi, M., Furukawa, Y., Matsuo, H., Fujiwara-Sakata, M. and Terano, Y. (1996) A neuropeptide "fulicin" regulates the contraction of a molluscan muscle. *Bull Hiroshima Pref Univ* 8:157–170.
27. Minakata, H., Ikeda, T., Muneoka, Y., Kobayashi, M. and Nomoto, K. (1993) WWamide-1, -2 and -3: novel neuromodulatory peptides isolated from ganglia of the African giant snail, *Achatina fulica*. *FEBS Lett* 323:104–108.
28. Ohta, N., Kanda, T., Kuroki, Y., Kubota, I., Muneoka, Y. and Kobayashi, M. (1990) Three novel tetradecapeptides isolated from the ganglia of the molluscs. In: N. Yanaihara (ed.): *Peptide Chemistry*. Protein Research Foundation, Osaka, 57–62.
29. Kobayashi, M., Fujita, K., Fujiwara-Sakata, M., Nomoto, K., Minakata, H. and Terano, Y. (1993) Structure-activity relations and the distribution of a neuropeptide containing a D-amino acid residue "fulicin" isolated from the African giant snail. In: N. Yanaihara (ed.): *Peptide Chemistry*. ESCOM, Leiden, 353–355.
30. Fujita, K., Minakata, H., Nomoto, K., Furukawa, Y. and Kobayashi, M. (1995) Structure-activity relations of fulicin, a peptide containing a D-amino acid residue. *Peptides* 16: 565–568.
31. Mor, A., Amiche, M. and Nicolas, P. (1992) Enter a new post-translational modification: D-amino acids in gene-encoded peptides. *Trends Biochem Sci* 17:481–485.
32. Kreil, G. (1994) Peptides containing a D-amino acid from frogs and molluscs. *J Biol Chem* 269:10967–10970.
33. Ikeda, T., Minakata, H., Fujita, T., Muneoka, Y., Kiss, T., Hiripi, L. and Nomoto, K. (1993) Neuropeptides isolated from *Helix pomatia*. Part 1. Peptides related to MIP, buccalin, myomodulin-CARP and SCP. In: N. Yanaihara (ed.): *Peptide Chemistry*. ESCOM, Leiden, 576–578.
34. Fujisawa, Y., Kubota, I., Ikeda, T., Minakata, H. and Muneoka, Y. (1991) A variety of *Mytilus* inhibitory peptides in the ABRM of *Mytilus edulis*: isolation and characterization. *Comp Biochem Physiol* 100C:525–531.
35. Ikeda, T., Yasuda-Kamatani, Y., Minakata, H., Kenny, P.T.M., Nomoto, K. and Muneoka, Y. (1992) *Mytilus*-inhibitory peptide analogues isolated from the ganglia of a pulmonate mollusc, *Achatina fulica*. *Comp Biochem Physiol* 101C:245–249.
36. Yasuda-Kamatani, Y. (1997) Significance of D-amino acids in novel neuropeptides from *Achatina fulica*. PhD thesis. Osaka University, Japan.

D-Amino Acids in Sequences of Secreted Peptides of Multicellular Organisms
ed. by P. Jollès

Opioid peptides from frog skin

M. Amiche, A. Delfour and P. Nicolas

Laboratoire de Bioactivation des Peptides, Institut Jacques Monod, 2 Place Jussieu, F-75251 Paris Cedex 05, France

Summary. The skin of the South American frogs *Phyllomedusa* secretes, in addition to numerous mammalian-like hormones and neuropeptides, several gene-encoded opioid peptides that contain a D-amino acid in position 2 of their sequence. Dermorphin, Tyr-D-Ala-Phe-Gly-Tyr-Pro-Ser-NH_2, dermenkephalin/deltorphin A, Tyr-D-Met-Phe-His-Leu-Met-Asp-NH_2 and the deltorphins, Tyr-D-Ala-Phe-Xaa-Val-Val-Gly-NH_2 (where Xaa is either Asp or Glu) are highly potent at, and exquisitely selective, for the μ- and δ-opioid receptors. D-Ala and D-Met present in dermorphin and related peptides are coded for by the usual codons in the corresponding messenger RNAs. Prepro-dermorphin/dermenkephalin and prepro-deltorphins have considerable sequence identities to precursors encoding 10–46-residue-long antimicrobial peptides – dermaseptins, brevinins, temporins, esculentins and gaegurins – originating from various amphibian species. The similarity between the prepro-regions of precursors encoding end products with strikingly different structures and biological activities supports the suggestion that the genes encoding these peptides are all members of the same family.

Introduction

One of the most fascinating aspects of comparative biochemistry is represented by the pioneering studies of Erspamer and colleagues on amphibian skin secretions [1, 2]. They identified in skin extracts some 30 different bioactive peptides bearing close structural similarities to mammalian peptides which act as hormones, neurotransmitters or neuromodulators, and sharing many of the same biological properties [3, 4]. As a result of the repeated discovery of structural correspondence between frog skin peptides and mammalian neuropeptides and hormones, Erspamer predicted that each peptide discovered in the skin would have a counterpart in the brain and gastrointestinal tract of mammals, i.e., the socalled brain-skin-gut triangle hypothesis [5, 6].

So far, more than 110 different biologically active peptides have been isolated from skin extracts at a still-increasing pace and with no signs yet of exhaustion [7]. As predicted by Erspamer, most of these peptides belong to families of biologically active peptides which have their counterparts in mammals. Examples are many and include caerulein/cholecystokinins, tachykinins, bradykinins, angiotensins, bombesin/gastrin-releasing peptide, hypophysiotropic neuropeptides and pancreatic polypeptide/peptide tyrosine-tyrosine/neuropeptide tyrosine [7]. Amphibian skin peptides are usually produced in such enormous quantities that it is possible to isolate enough material from a single specimen to determine the amino acid sequence and establish the pharmacological profile. As such, skin peptides represent a useful tool in the identification of mammalian neuropeptides that may be present in the brain in only minute quantities.

Among the peptide families so far described in amphibian skin, that of the opioid peptides is one of the most intriguing, as these peptides exhibit singular structural properties and outstanding biological activities.

The dermorphins, vertebrate opioid peptides which owe their biological activity to a D-amino acid

During the 1970s, research on mammalian endogenous ligands for opiate receptors led to the discovery of three distinct families of peptides, 5 to 31 residues long, termed enkephalins, endorphins and dynorphins [8] (Tab. 1). Although all these peptides contain the sequence Tyr-Gly-Gly-Phe at the N-terminal end, they arise from the processing of one of the three distinct polyprotein precursors, proenkephalin A [9], proopiomelanocortin [10] and prodynorphin [11]. Unfortunately, it was soon appreciated that mammalian opioid peptides hardly distinguish between the various types of

Table 1. Naturally occuring peptide opioids from mammalian brainand frog skin

Peptide	Amino acid sequence	Receptor selectivity		
		μ	δ	κ
Mammalian peptides				
Met-enkephalin	Y-G-G-F-M	++	+++	+
Leu-Enkephalin	Y-G-G-F-L	++	+++	+
β-endorphin	Y-G-G-F-M-T-S-E-K-S-Q-T-P-L-V-T L-F-K-N-A-I-I-K-N-A-H-K-K-G-Q	+++	+++	++
Dynorphin A (1–17)	Y-G-G-F-L-R-R-I-R-P-K-L-K-W-D-N-Q	+++	+	++++
Frog skin peptides				
Dermorphin[a]	Y-a-F-G-Y-P-S-amide	++++	+	
[Hyp[6]]-dermorphin	Y-a-F-G-Y-Hyp-S-Amide	++++	+	
[Trp[4], Asn[7]]-dermorphin	Y-a-F-W-Y-P-N	++++	+	
[Lys[7]]-dermorphin	Y-a-F-G-Y-P-K	++++	+	
[Trp[4], Ans[5]]-dermorphin-[1–5]	Y-a-F-W-N	++++	+	
Dermenkephalin[a] (Met-deltorphin)	Y-m-F-H-L-M-D amide	+	++++	
Ala-Deltorphin I	Y-a-F-D-V-V-G amide	+	++++	
Ala-Deltorphin II	Y-a-F-E-V-V-G amide	+	++++	
[D-Leu[5]]-deltorphin[b]	Y-L-F-A-D-V-S-T-I-G-D-F-F-H-S-I amide			

[a] D-amino acid residues are in lower case letters.

[b] This 17 amino acid peptide has been isolated from methanol extracts of the skin of the Brazilian frog, *Phyllomedusa burmeisteri* [47]. It displays a poor affinity for δ opioid receptors. The shorter amidated analogue [1–7] is virtually inactive on opioid binding sites in the periphery, but displays a clear-cut affinity for both δ- and μ-binding sites on rat brain membranes.

opioid receptors, making it difficult to explore the mechanism of opioid receptor selection for ligands and to define the individual role(s) played by the μ, δ and κ receptors in the brain and the pheriphery. Accordingly, considerable efforts have been devoted to the design and synthesis of various probes with improved affinity and selectivity at a single type of opioid receptor. Of the few type-selective ligands available to date, most are derived from the enkephalin sequence and include a D-amino acid residue in position 2 [12–14]. This fortuitous modification was found to produce analogues with prolonged bioactivity as a result of enhanced stability against proteolytic enzymes [15].

Soon after the discovery of the enkephalins, application of the principles laid down in the "brain-skin-gut triangle" hypothesis led Erspamer and collaborators to demonstrate the occurrence of large amounts of opioid activity in purified skin extracts from the South American hylid frogs, *Phyllomedusa sauvagii* and *P. rohdei* [16, 17]. However, the primary structures of the isolated opioid heptapeptide amides, respectively termed dermorphin (Tyr-D-Ala-Phe-Gly-Tyr-Pro-Ser-NH_2) and [Hyp^6]-dermorphin Tyr-D-Ala-Phe-Gly-Tyr-Hyp-Ser-NH_2), are so drastically different from those of the mammalian opioids that it is very unlikely that the enkephalins, endorphins and dynorphins represent the mammalian counterparts of the dermorphins. Also, dermorphin molecules were the first examples of biologically active peptides originating from a nonmicrobial source to contain a D-amino acid residue, the D configuration being absolutely essential for their pharmacological activities [16, 17]. Last but not least, extensive *in vitro* and *in vivo* studies have unequivocally proven dermorphins to be the most potent and selective naturally occurring opioids for μ receptors in the central nervous system and in the periphery [18]. In receptor-binding experiments, dermorphin was shown to bind almost exclusively to the μ-receptor, being more potent and selective than DAGO (Tyr-D-Ala-Met-Phe-Gly-ol), a prototypical μ-selective probe. The pharmacological and the behavioural effects of dermorphin mimic, with exaggeration, those of morphine. For instance, dermorphin is about 1000 times more potent than morphine in producing long-lasting analgesia and in inducing catalepsy upon intracerebroventricular administration in rodents. β-endorphin, the most potent mammalian opioid peptide, shows barely 5–10% of the activity of dermorphin. Dermorphin also produces considerable analgesia in rodents by subcutaneous (SC) and intravenous (IV) injections, being 3–5 times more potent than morphine. However, the association of antinociception, tolerance and dependence observed in all other opioids and alkaloid opiates is also valid for dermorphin [18].

Through a combination of peptide chemistry and complementary DNA (cDNA) cloning (*vide infra*), several other members of the dermorphin family, i.e., μ-selective D-amino acid-containing ligand, were discovered later from the same amphibian sources [19] (Tab. 1). For the present, the dermorphin subfamily is composed of six members five heptapeptides and one pentapeptide, all containing a D-Ala residue in position 2 and all being highly potent at, and selective for, the μ-opioid receptors. The search for peptides of this type in the skin of other frogs was not successful. To date, dermorphins have only been detected in the skin of Phyllomedusinae from South and Middle America [18].

Demorphins are gene-encoded: A new posttranslational modification

The discovery of dermorphin raised a major question about its biosynthesis, since no D-amino acid-containing peptide originating from the eukaryotic world was yet known. Instead, multienzymatic synthesis of peptides that contain D-amino acids was regarded as purely a prokaryotic function. Since eukaryotic cells invariably use only the L-amino acid pool and ribosome-mediated protein synthesis, the presence of a D-amino acid peptide in the cells of vertebrate animals was thus received with scepticism. On the other hand, it was well established that all the known carboxamidated neuropeptides and hormones arise from processing of ribosomally made precursor molecules in which the progenitor peptide sequence is immediately followed by a Gly residue involved in the formation of the C-terminal peptide amide. Since dermorphin is carboxamidated, this was clear evidence in favour of ribosomal synthesis. The cloning of at least two biosynthetic precursors of dermorphin from *P. sauvagii* skin messenger RNA (mRNA) partially resolved the riddle of its formation by demonstrating that dermorphin is indeed a product of ribosomal synthesis [20]. Moreover, the structure of the prepro-dermorphin is consistent with that expected for classical prohormones (Fig. 1): it contains a hydrophobic signal peptide at its amino terminus, which is followed with repetitive segments of about 35 amino acid residues. Each of these segments codes for a copy of the dermorphin progenitor sequence, which contians an additionnal Gly residue at its carboxyl terminus; this extra residue is required for the carboxamidation of the mature heptapeptide. The progenitor sequences are also bracketed by typical bibasic cleavage signals for prohormone convertases.

Unlike D-Ala-containing lantibiotics, such as nisin and epidermin, for which the incorporation of a D-Ala in the mature peptides results from posttranslational modifications of L-Ser, the D-Ala present in dermorphin is coded for by the usual L-Ala codon (GCG). The most plausible scheme for this transformation would involve a classical synthesis immediately followed by a posttranslational L-D isomerization of the L-Ala residue [21]. However, the precise mechanism of the reaction(s) as well as the participating enzyme(s) remain to be discovered.

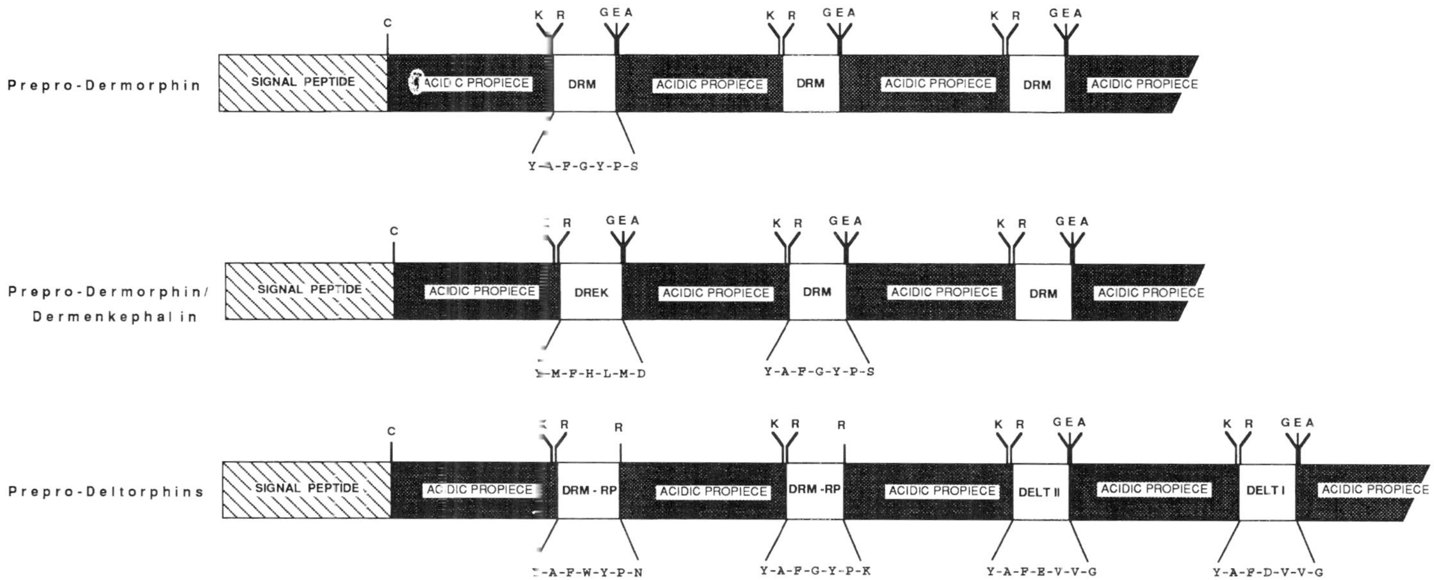

Figure 1. Block diagrams of propro-dermorphin and prepro-dermorphin/dermenkephalin cDNAs from *P. sauvagii* skin [20] and prepro-deltorphins cDNA from *P. bicolor* skin [19]. Regions of great similarity among all three cDNAs are labeled with the same boxed patterns. C, cysteine; K, lysine; R, arginine; GEA, Gly-Glu-Ala tripeptide; DRM, dermorphin; DREK, dermenkephalin; DRM-RP, dermorphin-related peptide; DELT, deltorphin. Only partial sequences of the cDNAs encoding prepro-dermorphins and prepro-deltorphins are shown, since the number of repeats varies and is clone-dependent.

The deltorphins, highly selective δ agonists

The two cloned prodermorphins from *P. sauvagii* are nearly identical except that one has a distinct predicted heptapeptide sequence, Tyr-Met-Phe-His-Leu-Met-Asp, instead of the first dermorphin sequence (Fig. 1). Assuming the same processing pathway as for the release of dermorphin, processing of such a precursor was expected to yield, beside four copies of dermorphin, a new carboxamidated peptide, Tyr-D-Met-Phe-His-Leu-Met-Asp-NH_2, with unknown biological activities. The predicted peptide, termed deltorphin A [22], dermenkephalin [23] or dermorphin gene-associated peptide [24] was synthesized and was independently shown by three groups to be a highly specific ligand for the δ-opioid receptor. In contrast, the L-Met^2 isomer was consistently found to be inactive. This putative D-Met-containing opioid peptide was indeed isolated from skin extracts of *P. sauvagii* [22, 25].

In addition, two closely related opioid heptapeptides were isolated from the skin of another arboreal frog species, *P. bicolor*. These peptides were termed Ala-deltorphin I (Tyr-D-Ala-Phe-Asp-Val-Val-Gly-NH_2) and Ala-deltorphin II (Tyr-D-Ala-Phe-Glu-Val-Val-Gly-NH_2), both of which contain a D-Ala residue in position 2 [26]. The concentration of these peptides in the skin of *P. bicolor* was enormous. For instance, as much as 1300 μg/g of Ala-deltorphin I was found in a single specimen of dried skin [18]. Although these peptides share with dermorphin a common tripeptide amino end, Tyr-D-Ala-Phe, they proved to be formidable agonists for the δ-opioid receptors, being even more potent and selective than dermenkephalin (deltorphin A). This clearly emerged from binding assays on isolated preparations and was confirmed in the binding assay on rat brain membranes. Once again, the L-Ala^2 forms of these peptides were found to be completely inactive.

Again, cDNA cloning techniques led to the identification of two precursors (Fig. 1) containing not only the progenitor sequences of the two new deltorphins but also the sequences of three novel dermorphins, i.e., Tyr-D-Ala-Phe-Trp-Tyr-Pro-Asn, Tyr-D-Ala-Phe-Gly-Tyr-Pro-Lys and Tyr-D-Ala-Phe-Trp-Asn (Tab. 1), which were further isolated from skin extracts [19]. Whereas in the precursors the C-terminal Gly residue of Ala-deltorphins I and II are followed by an extra Gly residue, this is not the case for the additional peptides related to dermorphin. Once again, the D-Ala in the mature peptides was found to be coded by the usual L-aa codon (GCG). Thus, pro-dermorphin/dermenkephalin and pro-deltorphins offer unique examples of plurifunctional biosynthetic precursors which generate two types of biologically active peptides, dermorphins and dermenkephalin/deltorphins, each of which contains a D-amino acid residue and is specific for a single subtype of the opioid receptor.

Dermorphins and deltorphins as models for investigating the mechanism of opioid receptor selection for ligands

The discovery of dermorphins and dermenkephalin/deltorphins provided tractable tools for understanding the yet unknown mechanism of opioid receptor selection for peptide ligands. Indeed, despite a common N-terminal tripeptide, Tyr-D-Xaa-Phe (where Xaa is Met or Ala), these two groups of peptides exhibit enormous differences in receptor selectivities but still bind to their own receptors with similar affinities. It was thus hypothesized that the amino and carbocyl terminal domains of these peptides represented a consensus "receptor-binding pharmacophore" and a specific "receptor address domain", respectively, i.e., determinants imparting opposite selectivities of dermorphin and dermenkephalin/deltorphins were to be traced to different constitution/conformation of their respective C-terminal tails. This was first nicely evidenced by the damatic reversal of receptor selectivity for $\mu > \delta$ to $\delta > \mu$ following replacement of the C-terminal tri- or tetrapeptide of dermorphin with the dermenkephalin counterparts [27, 28]. Also, it was found that the tetrapeptide amino ends of dermenkephalin and Ala-deltorphins I and II showed high preference for μ receptors over δ receptors. Identification of common determinants concurring to the remarkably efficient targeting of dermenkephalin and detorphins towards the δ receptors was then achieved through structure-activity relationship studies carried out on an extensive series of point-substituted synthetic analogues. Results showed that the C-terminal domain confers high δ selectivity to the ligand by two modes: increased affinity for δ sites and rejection of dermenkephalin/delorphin peptides by the μ sites [18, 29, 30]. Briefly, the presence of anionic side chains within the C-terminus of dermenkephalin and deltorphins enhances δ selectivity mostly by preventing μ binding. Position 4 contributes to δ selectivity by inhibiting high-affinity binding to μ sites. In contrast, hydrophobic aliphatic side chains at positon 5 and, to a lesser extent position 6, enhance δ affinity.

Various conformational models have been proposed in an attempt to elucidate the precise mechanism by which the C-terminal addresses of dermenkephalin and deltorphins discriminate between opioid-receptor types. Conformational analysis based either on internal energy calculations or combined use of quantitative two-dimensional nuclear magnetic resonance (2D NMR) spectroscopy in DMSO and cryoprotective mixtures and energy calculations [31–39] suggested that dermorphin has a linear and flexible C-terminus, while that of dermenkephalin/deltorphins comes into close contact with the tripeptide amino end. Thus, the C-terminus of δ-selective peptide opioids was thought to confer selectivity towards the δ receptor either by anchoring the Tyr^1 and Phe^3 side chains in an orientation that is best suited for the δ-binding site, or fashioning a bulky structure which interferes sterically with the complementarity between these peptides and the μ-binding site. The more recent studies favour the first hypothesis, initially proposed by Hruby and co-workers, whose major tenet is that Tyr[1] and Phe[3] moieties interact with similar complementary binding sites on the μ- and δ-opioid receptors but in different orientations. The C-terminus of dermenkephalin/deltorphins would thus act as a scaffold to anchor and coordinate the orientations of the aromatic rings that would be selectively recognized by the δ receptor.

On the other hand, dermenkephalin and the deltorphins display very similar N-termini but strikingly different C-terminal tails. Nevertheless, both types of peptides are endowed with similar potencies and selectivities towards the δ receptors. This suggests that conformational attributes common to dermenkephain and deltorphins underlie their ability to be selectively recognized by the δ sites. Indeed, recent experiments support this proposal by showing that dermenkephalin and deltorphins exhibit unifying conformational features despite differing constitutions of their carboxy ends: a common S-shaped arrangement in the peptide N-termini and similar orientations of the main pharmacophoric elements Tyr[1] (t) and Phe[3] (g-), and may be t for Xaa[4] [39]. This common stuctural trend adds further support to the proposal that δ- and μ-selective opioid peptides from frog skin differ only in presence or absence of a spatial overlap of the N- and C-terminal domains.

A gene family encoding opioid peptides and antimicrobial peptides originating from various frog species

Aside numerous mammalian-like neuropeptides and hormones, frog skin synthesizes and expels several antimicrobial cationic peptides thought to be involved in the defence of the naked skin against microbial invasion, and to aid in wound repair [40]. To date, microbicidal peptides of the amphibian skin may be grouped into three broad families (Tab. 2): (i) linear amphipathic helix-forming peptides such as the dermaseptins S and B from the arboreal frogs *P. sauvagii* and *P. bicolor*, respectively; (ii) four different groups of related peptides isolated from various pecies of the Ranidae family, namely the brevinins, the gaegurins and the esculentins, which all contain two cysteinyl residues at the C-terminus linked in a disulfide bridge; and (iii) the temporins, isolated from *Rana temporaria*, which are the smallest antimicrobial peptides known so far, being only 10–13 residues in length.

Isolation and characterization of cDNA-encoding precursors of dermaseptins B from *P. bicolor* [41] showed that the nucleotide and deduced amino acid sequences of prepro-dermaseptins are strikingly similar to those of prepro-dermorphin/dermenkephalin and prepro-deltorphins. As shown in Figure 2, the four precursors are composed of distincts domain arranged in a similar manner. The amino-terminal domain includes a 22-residue signal peptide and is followed by an acidic domain containing 20–24 residues.

Table 2. Some prototypical members of families of antimicrobial peptides from frog skin

Peptide	Amino acid sequence
Dermaseptin B1	AMWKDVLKKIGTVALHAGKAALGAVADTISQ ($CONH_2$)
Dermaseptin B2	GLWSKIKEVGKEAAKAAAKAAGKAALGAVSEAV ($CONH_2$)[a]
Dermaseptin B3	ALWKNMLKGIGKLAGQAALGAVKTLVGA
Dermaseptin B4	ALWKDILKNVGKAAGAVLNTVTDMVNQ ($CONH_2$)
Dermaseptin B5	GLWNKIKEAAKSAGKAALGFVNEMV ($CONH_2$)
Dermaseptin B6	ALWKDILKNAGKAALNEINQLVNQ ($CONH_2$)
Brevinin-1E	FLPLLAGLAANFLPKIF**C**KITRK**C**[b]
Esculentin-1	GIFSKLAGKKLKNLLISGLKNVGKEVGMDVVRTGIDIAG**C**KIKGE**C**
Gaegurin-4	GILDTLKQFAKGVGKDLVKGAAQGVLSTVS**C**KLAKT**C**
Temporin H	LSPNLLKSLL ($CONH_2$)

[a] Initially isolated Daly and co-workers and termed adenoregulin [48].
[b] The cysteinyl residues which are linked in a disulfide bridge are in bold.

The third domain consists of a progenitor sequence of variable-length coding for either an antimicrobial or opioid peptide that is flanked by a pair of basic residues at its amino end and a tripeptide (Gly-Glu-Ala/Gln) at its C-terminus. The Gly residue is involved in the formation of the C-terminal peptide amide in all four precursors. Whereas prepro-dermaseptins each contain only a single copy of an acidic propiece and a bioactive sequence assembled in tandem, prepro-dermorphin/dermenkephalin and prepro-deltorphins are made up of several highly homologous repeats of this tandem module. A comparison of the cDNA and deduced amino acid sequences of the four prepro-proteins shows that the signal peptides (72% identical at the amino acid level) and the acidic propieces (50% identical at the amino acid level) are very similar. The putative processing sites of the precursors are also arranged identically; the paired basic residues Lys-Arg and the C-terminal tripeptide extension Gly-Glu-Ala/Gln that flank the biactive sequence(s) are the same in the four prepro-proteins. This similarity also extends into the 5′-untranslated portions of the respective mRNA (71% nucleotide positional identity). However, the similarity is lower in the 3′-untranslated regions of the mRNAs.

Unexpected similarities were also found between the conserved prepro-regions of prepro-dermaseptins and prepro-dermorphin/dermenkephalin/deltorphins from *Phyllomedusa* spp. and those of precursors-encoding brevinins, esculentins, gaegurins and temporins from *Rana* spp., respectively [42–44] (Fig. 2). These 10–46-residue-long peptides are synthesized as part of precursor proteins whose amino terminal domain includes a signal peptide of 22 residues, followed by an acidic propiece containing 16–25 residues, a paired basic residue Lys-Arg and a progenitor antimicrobial sequence. The extensive similarities between the prepro-regions of precursors originating from various amphibian species and encoding end products with strikingly different structure and biological activity support the suggestion that the genes encoding these peptides are all members

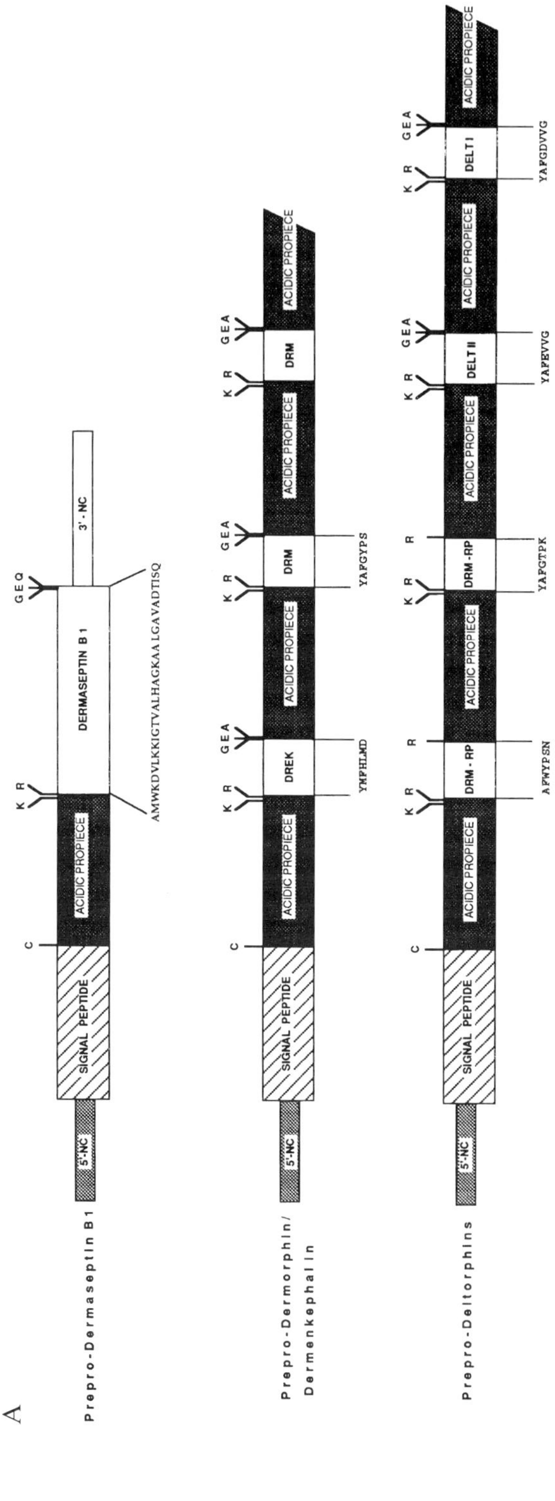
A
Prepro-Dermaseptin B1
5'-NC
SIGNAL PEPTIDE
C
ACIDIC PROPIECE
K R
DERMASEPTIN B1
G E Q
3'-NC
AMWKDVLKKIGTVALHAGKAALGAVADTISQ
Prepro-Dermorphin/
Dermenkephalin
5'-NC
SIGNAL PEPTIDE
C
ACIDIC PROPIECE
K R
DREK
G E A
YMFHLMD
ACIDIC PROPIECE
K R
DRM
G E A
YAFGYPS
ACIDIC PROPIECE
K R
DRM
G E A
ACIDIC PROPIECE
Prepro-Deltorphins
5'-NC
SIGNAL PEPTIDE
C
ACIDIC PROPIECE
K R
DRM - RP
R
APWYPSN
ACIDIC PROPIECE
K R
DRM - RP
R
YAFGTPK
ACIDIC PROPIECE
K R
DELT II
G E A
YAFEVVG
ACIDIC PROPIECE
K R
DELT I
G E A
YAFGDVVG
ACIDIC PROPIECE

Figure 2A

B

Precursor	Signal peptide																						Acidic propiece		
	1									10										20					
Dermaseptins B1	M	D	I	L	K	K	S	L	F	L	V	L	F	L	G	L	V	S	L	S	I	C	**E**	**E**	**E**
Dermorphin	M	S	F	L	K	K	S	L	L	L	I	L	F	L	G	L	V	S	L	S	V	C	**K**	**E**	**E**
Deltorphins	M	S	F	L	K	K	S	L	L	L	V	L	F	L	G	L	V	S	H	S	V	C	**K**	**E**	**E**
Brevinin-2E	M	F	T	**M**	K	K	S	L	L	L	I	F	F	L	G	T	I	S	L	S	L	C	**Q**	**E**	**E**
Temporin H	M	F	T	L	K	K	S	L	L	L	L	F	F	L	G	T	I	N	L	S	L	C	**E**	**E**	**E**
Gaegurin-5	M	F	T	L	K	K	S	L	L	L	L	F	F	L	G	T	I	S	L	S	L	C	**E**	**E**	**E**
	*				*	*	*	*		*			*	*	*					*		*		*	*

Figure 2. (A) Block diagrams of the prepro-dermaseptin B1 cDNA [41], prepro-dermorphin/dermenkephalin [20] and prepro-deltorphins [19] cDNAs. Regions of great similairty among all three cDNAs are labeled with the same boxed paterns as follows: 5′ NC, 5′-noncoding region (dark dotted box); signal peptide (hatched box); acidic spacer peptide (black box); C, cysteine; KR, Lys-Arg dipeptide; GEQ/A, Gly-Glu-Gln/Ala tripeptide. Open boxes denote sequences with low or no homology among the four precursors: 3′ NC, 3′-noncoding region; DRM, dermorphin; DREK, dermenkephalin; DRM-RP, dermorphin-related peptides; DELT, deltorphin. Only partial sequences of the cDNAs encoding prepro-dermorphin and prepro-deltorphins are shown, since the number of repeats varies and is clone dependent. (B) Sequence alignments of te predicted amino acid sequences of the signal peptides from prepro-dermaseptin B1 [41], prepro-dermorphin [20], prepro-deltorphins [19], prepro-brevinin-2E (BREV-2E) [42], prepro-temporin H (TEMPO H [43] and prepro-gaegurin 5 (GAEG-5) [44]. Amino acids corresponding to the beginning of the acidic propiece are shown in bold letters. Identical (*) residues among sequences are highlighted.

A

ttttgtgcataaagtgtacaacattaa
ttactgtgcctttctgcacaaaacatccaaaacacatcaaaaagaaacaaaaaa
actcagttttctggtgctttttgatgcatttttttgtgtattttaaatattt
attaacattgtaacaattaaggagagctaaattatcattgtcaatctgt**ccaat**
actggctatttttccttatgttttatac**taacaaaa**catttatttccattta
agaagagaaatacaaacctgtttaaccaatataaatcttttgacagaccaaac
ATGGCTTTCCTGAAGAAATCTCTTTTCCTTGTACTGTTCCTTGGATTGGTCTCT
M A F L K K S L F L V L F L G L V S
CTTTCTATTTGTGAAGAAGA [gtaagtactaatatattataatgaaacaatga
L S I C E E E
atctgggttttcattctttgctgtgataaggggctgggcacagctcagagctgt
aggaaaaagtatcctaaatttaaaaaaaaaaattatattgcctcctaacag]**G**
AAAAGAGAAAATGAAGATGAGGAGGAACAAGAAGATGACGAGCAAAGTGAAATG
K R E N E D E E E Q E D D E Q S E M
AAGAGAGGACTGTGGAGTAAAATAAAAGAAGCAGGAAAAGCGGCATTAACAGCT
K R G L W S K I K E A G K A A L T A
==
GCAGGAAAAGCAGCTTTAGGTGCAGTTTCTGATGCGGTGGGAGAGCAAtaaagt
A G K A A L G A V S D A V G E Q ***
=====================================
tagtaaaatgtaaagtcaatcaaattgctttgaggagcacagttataaataatt
atgtcaaatcttt**attaaaa**taaataaaaatgatatgactgtctgtcctttta
catacacattttatatacatttatgttttgtgtatacataattaagattataga
taaaaggtaagagcactaaaatactgaagtatagggaatcctaaaacgagaatg
gagaaaacaaatgtgtggggatataagaagaataaaatttagttcaaaaggaa
agtatatgatagtcatgagtataattgtatacacatccagaa

B

Figure 3. Structure of a prepro-dermaseptin gene (gene *Drg2*). (A) Nucleotide sequences of coding exons are indicated by capital letters. The intronic sequence, 5′- and 3′-untranslated sequences are shown in lowercase letters. The putative TATA-box, CAAT box and polyadenylation signal are in bold letters. The predicted mature prepro-dermaseptin sequence is shown in bold letters under the coding DNA. The predicted signal peptide is underlined, and the mature dermaseptin sequence is doubly underlined. The stop codon is indicated by asterisks. (B) Schematic representation of the *Drg2* gene. Coding exons are drawn as rectangle, the intron as a thick line.

of the same family. This hypothesis received further support through cloning of several genes encoding dermaseptins B [45, 46]. As depicted in Figure 3, most of these genes have a two-exon coding structure interrupted by a small intron, wherein exon 1 encodes the 22-residue hydrophobic signal peptide and the first three amino acids of the acidic propiece; exon 2 contains the 18 additional acidic residues of the propiece plus the processing signal Lys-Arg and a dermaseptin progenitor sequence. Coding exon 1 shares 48% nucleotide identity (50% at the amino acid level) with the corresponding regions of the esculentin, brevinin, temporin, gaegurin, dermorphin/dermenkephalin and deltorphin cDNAs. Also, the 5′-region in the gene is ~60% identical to the 5′-untranslated regions of the cDNAs; at least over the first 13 bp immediately upstream of the ATG initiation codon; beyond that point, sequences diverge. This suggests that opioid and antimicrobial gene subfamilies have evolved with a conserved "secretory cassette" exon followed by exon(s) encoding end products with unrelated sequences and activities. The duplication and recombinational events that promoted the association of such a homologous secretory exon with nucleotide sequences coding for a variety of end products in various frog species remain to be elucidated but are likely to have occurred at the very early stages of evolution.

D-amino acid-containing opioid peptides in mammals: An open question

The concept of the brain-skin-gut triangle proposed by Erspamer states that peptides occuring in amphibian skin exist in a similar, if not identical, sequence in mammalian brain and gut. Since it is hardly probable that the enkephalins, endorphins and dynorphins represent the mammalian counterparts of dermorphin, dermenkephalin and deltorphins, several groups have undertaken an active search for these peptides in mammalian tissues. By using antibodies specific for different part of dermorphin, dermenkephalin and deltorphins or their respective precursor proteins, small amounts of cross-reacting materials that behave like the authentic peptides by chromatographic comparison with synthetic standards were repetitively detected in various brain areas, the retina, the olfactory bulb and the gastrointestinal tract [18]. Although these results bring arguments in favour of an endogenous synthesis and processing of pro-dermorphin/dermenkephalin and pro-deltorphins in mammalian tisues, the exact chemical nature of these immunoreactive substances has never been established.

References

1. Erspamer, V. (1973) Biogenic amines and active polypeptides of the amphibian skin. *Ann Rev Pharmacol* 11:327–350.
2. Erspamer, V. (1984) Half a century of comparative research on biogenic amines and active peptides in molluscan tissues and amphibian skin. *Comp Biochem Physiol* 79C:1–7.
3. Erspamer, V., Melchiori, P., Falconieri-Erspamer, G., Montecucchi P. and De Castiglione, R. (1985) Phyllomedusa skin: a huge factory and store-house for a variety of active peptides. *Peptides* 6:7–12.
4. Erspamer, V., Falconieri-Erspamer, G. and Cei, J.M. (1986) Active peptides in the skin of 230 American amphibian species. *Comp Biochem Physiol* 85C:125–137.
5. Erspamer, V. and Melchiorri, P. (1980) Active polypeptides from the amphibian skin to gastrointestinal tract and brain of mammals (1980) *Trends Pharmacol Sci* 1:391–395.

6. Erspamer, V., Melchiorri, P., Broccardo, M., Falconieri-Erspamer, G., Falaschi, P., Impota, G., Negri, L. and Renda, T. (1981) The brain-skin-gut triange: new peptides. *Peptides* 2: 7–16.
7. Lazarus, L.H. and Attila, M. (1993) The toad, ugly and venomous, wears yet a precious jewel in his skin. *Prog Neurobiol* 41:473–507.
8. Kosterlitz, H.W. (1985) Opioid peptides and their receptors. *Proc R Soc B* 225:27–40.
9. Noda, M., Furutani, Y., Takahashi, M., Toyosato, M., Hirose, T., Inayama, S., Nakanishi, S. and Numa, S. (1982) Cloning and sequence analysis of cDNA for bovine adrenal pre-proenkephalin. *Nature* 295:205–206.
10. Nakanishi, S., Inoue, A., Kita, T., Chang, A.C.Y., Nakamura, M., Cohen, S. and Numa, S. (1979) Nucleotide sequence of coned cDNA for bovine corticotropin-β lipotropin recursor. *Nature* 278:423–427.
11. Kakidani, H., Furutani, Y., Takahashi, M., Noda, M., Morimoti, Y., Hirose, T., Asai, M., Inayama, S., Nakanishi, S. and Numa, S. (1982) Cloning and sequencing of cDNA for porcine βEP/dynorphin precursor. *Nature* 298:245–250.
12. Kosterlitz, H.W. and Paterson, S.J. (1981) Tyr-D-Ala-Gly-MetPhe-NH$(CH_2)_2$-OH is a selective ligand for the μ-opiate binding site. *Br J Pharmacol* 73:299 P.
13. Mosberg, H., Hurst, R., Hruby, V.J., Gee, K., Yamamura, I., Galligan, J.J. and Burks, T.F. (1983) Bis-penicillamine enkephalins possess highly improved specificity toward delta opioid receptors. *Proc Natl Acad Sci USA* 80:5871–5874.
14. Delay-Goyet, P., Seguin, C., Gacel, G. and Roques, B.P. (1988) (^{3}H)(Ser^2 (*O-tert*-butyl) Leu^5) enkephalyl-Thr^6 and (D-Ser^2 (*O-tert*-butyl) Leu^5) enkephalyl-Thr^6/(*O-tert*-butyl), 2 new enkepahlin analogs with both a good selectivity and a high affinity toward delta opioid binding sites. *J Biol Chem* 263:4124–4130.
15. Pert, C.B., Pert, A., Chang, J.K. and Fong, B.T.N. (1976) (D-Ala^2)-Met-enkephainamide, a potent, long-lasting synthetic pentapeptide analgesic. *Science* 194:330–332.
16. Montecucchi, P.C., De Castiglione, R., Piani, S., Gozzini, L. and Erspamer, V. (1981) Amino acid composition and sequence of dermorphin, a nove opiate-ike peptide from the skin of *Phyllomedusa sauvagei*. *Int J Peptide Protein Res* 17:275–283.
17. Montecucchi, P., De Castiglione, R. and Erspamer, V. (1981) Identification of dermorphin and Hyp dermorphin in skin extracts of the Brazilian frog *Phyllomedusa rodhei*. *Int J Peptide Protein Res* 17:316–321.
18. Erspamer, V. (1992) The opioid peptides of the amphibian skin. *Int J Dev Neurosci* 10:3–30.
19. Richter, K., Egger, R., Negri, L., Severini, C. and Kreil, G. (1990) cDNAs encoding (D-Ala^2) deltorphin precursors from skin of *Phyllomedusa bicolar* also contain genetic information for 3 dermorphin-related opioid peptides. *Proc Natl Acad Sci USA* 87:4836–4839.
20. Richter, K., Egger, R. and Kreil, G. (1987) D-alanine in the frog skin peptide dermorphin is derived from L-alanine in the precursors. *Science* 238:200–202.
21. Mor, A., Amiche, M. and Nicolas, P. (1992) Enter a new post-translational modification: D-amino acids in gene-encoded peptides. *Trends Biochem Sci* 17:481–485.
22. Kreil, G., Barra, D., Simmaco, M., Erspamer, V., Falconieri-Ersparmer, G., Negri, L., Severini, C., Corsi, R. and Melchiorri, P. (1989) Deltorphin, a novel amphibian skin peptide with high affinity and selectivity for the delta opioid receptor. *Eur J Pharmacol* 162:123–128.
23. Amiche, M., Sagan, S., Mor, A., Delfour, A. and Nicolas, P. (1989) Dermenkephalin, a potent and fully specific agonist for the delta opioid receptor. *Mol Pharmacol* 35:774–779.
24. Lazarus, L.H., Wilson, W.E., De Castiglione, R. and Guglietta, A. (1989) Dermorphin gene sequence peptide with high affinity and selectivity for delta opioid receptors. *J Biol Chem* 264:3047–3050.
25. Mor, A., Delfour, A., Sagan, S., Amiche, M., Pradelles, P., Rossier, J. and Nicolas, P. (1989) Isolation of dermenkephalin from amphibian skin, a high-affinity delta selective opioid heptapeptide containing a D-amino acid residue. *FEBS Lett* 255:269–274.
26. Erspamer, V., Melchiorri, P., Falconieri-Erspamer, G., Negri, L., Corsi, R., Severini, C., Barra, D., Simmaco, M. and Kreil, G. (1989) Deltorphins: a family of naturally occuring peptides with high affinity and selectivity for the delta opioid binding sites. *Proc Natl Acad Sci USA* 86:5188–5192.
27. Sagan, S., Amiche, M., Delfour, A., Mor, A., Camus, A., Nicolas, P. (1989) Molecular determinants of receptor affinity and selectivity of the natural delta opioid agonist, dermenkephalin. *J Biol Chem* 264:17100–17106.

28. Sagan, S., Amiche, M., Delfour, A., Camus, A., Mor, A., Nicolas, P. (1989) Differential contributions of C-terminal regions of dermorphin and dermenkephalin to opioid site selectivity and binding potency. *Biochem Biophys Res Commun* 169:726–732.
29. Lazarus, L.H., Bryant, S.D., Attila, M. and Salvadori, S. (1994) Frog skin opioid peptides: a case for environmental mimicry. *Environ Health Perspect* 102:648–654.
30. Charpentier, S., Sagan, S., Amiche, M., Naïm, M., Delfour, A. and Nicolas, P. (1994) Mechanism of opioid receptor selection by the address domain of dermenkephalin. *Eur J Pharmacol* 266:175–180.
31. Temussi, P.A., Picone, D., Tancredi, T., Tomatis, R., Salvadori, S., Marastoni, M. and Balboni, G. (1989) Conformational properties of deltorphin: new features of the delta opioid receptor. *FEBS Lett* 247:283–288.
32. Balboni, G., Marastoni, M., Picone, D., Salvadori, S., Tancredi, T., Temussi, P.A. and Tomatis, R. (1990) New features of the delta opioid receptor: conformational properties of deltorphin I analogues. *Biochem Biophys Res Commun* 169:607–612.
33. Amadeo, P., Motta, A., Tancredi, P., Salvadori, S., Tomatis, R., Picone, D., Saviano, G. and Temussi, P.A. (1992) Solution structure of deltorphin I at 265 K: a quantitative NMR study. *Peptide Res* 5:48–55.
34. Nikoforovich, G.V., Prakash, O., Gehrig, C. and Hruby, V.J. (1993) Conformation of the dermenkepahlin backbone in DMSO solution by a new approach to the solution conformation of flexible peptides. *J Am Chem Soc* 115:3399–3406.
35. Naïm, M., Charpentier, S., Nicolas, P. and Baron, D. (1993) Quantitative 2-D NMR study of dermenkephalin, a highly potent and selective delta opioid peptide. *Biopolymers* 33: 1889–1900.
36. Duchesne, D., Naïm, M., Nicolas, P. and Baron, D. (1993) Folding trends in a flexible peptide. *Biochem Biophys Res Commun* 195:630–636.
37. Ohno, Y., Segawa, M., Ohishi, H., Doi, M., Kitamura, K., Ishida, T., Inoue, M. and Shita, T. (1993) Conformation of deltorphin II in membrane environment studied by 2-D NMR spectroscopy and molecular dynamic calculations. *Eur J Biochem* 212:185–191.
38. Naïm, M., Nicolas, P. and Baron, D. (1996) Folded conformations of the delta selective opioid dermenkephalin with head-to-tail interaction. *Int J Peptide Protein Res* 47:353–360.
39. Naïm, M., Nicolas, P., Benajiba, A. and Baron, D. (1998) Solution conformations of deltorphin I issued from combined used of quantitative 2D-NMR and energy calculations. A comparison with dermenkephalin. *J Peptide Res*; *in press*.
40. Nicolas, P. and Mor, A. (1995) Peptides as weapon against microorganisms in the chemical defense system of vertebrates. *Ann Rev Microbiol* 49:277–304.
41. Amiche, M., Ducancel, F., Mor, A., Boulain, J.C., Menez, A. and Nicolas, P. (1994) Precursors of vertebrate peptide antibiotics dermaseptin B and adenoregulin have extensive sequence identities with precursors of opioid peptides dermorphin, dermenkephalin and deltorphins. *J Biol Chem* 269:17847–17852.
42. Simmaco, M., Mignogna, G., Barra, D. and Bossa, F. (1994) Antimicrobial peptides from skin secretions of *Rana esculenta*. *J Biol Chem* 269:11956–11961.
43. Simmaco, M., Mignogna, G., Canofeni, S., Miele, R., Mangone, M.L. and Barra, D. (1996) Temporins, antimicrobial peptides from the european red frog *Rana temporaria*. *Eur J Biochem* 242:788–792.
44. Park, J.M., Lee, J.Y., Moon, H.M. and Lee, B.J. (1995) Molecular cloning of cDNAs encoding precursors of the frog skin antimicrobial peptides from *Rana rugosa*. *Biochim Biophys Acta* 1264:23–25.
45. Vouille, V., Amiche, M. and Nicolas, P. (1997) Structure of genes for dermaseptins B, antimicrobial peptides from frog skin. *FEBS Lett* 414:27–32.
46. Fleury, Y., Vouille, V., Beven, L., Amiche, M., Woblesky, H., Delfour, A. and Nicolas, P. (1998) Synthesis, antimicrobial activity and gene structure of a novel member of the dermaseptin B family. *Biochim Biophys Acta* 1396:228–236.
47. Barra, D., Mignogna, G., Simmaco, M., Pucci, P., Severini, C., Falconieri-Erspamer, G., Negri, L. and Erspamer, V. (1994) (D-Leu 2) deltorphin, a 17 amino acid opioid peptide from the skin of the Brazilian hylid frog *Phyllomedusa burmeisteri*. *Peptides* 15:199–202.
48. Daly, W.J., Caceres, J., Moni, W.R., Gusovski, F., Moos, M., Seamon, B.K., Milton, K. and Myers, W.C. (1992) Frog secretions and hunting magic in the upper amazon: identification of a peptide that interacts with an adenosine receptor. *Proc Natl Acad Sci USA* 89:10960–10963.

D-Amino Acids in Sequences of Secreted Peptides of Multicellular Organisms
ed. by P. Jollès

D-Amino acids in crustacean hyperglycemic neurohormones

A. Huberman and M.B. Aguilar

Instituto Nacional de la Nutrición "Salvador Zubirán", Department of Biochemistry, Vasco de Quiroga 15, Tlalpan, Mexico City 14000, Mexico

History

In 1944, Abramowitz et al. [1] described a "diabetogenic factor" found in aqueous extracts of the crab *Uca pugilator* that when injected into the blue crab, *Callinectes sapidus*, produced an intense hyperglycemia in a very short time. Most of this activity was found in the sinus gland of the eyestalk, was effective over a wide dilution range and was interspecific. Kleinholz et al. [2, 3] were able to show in the crab *Libinia emarginata* that removal of both eyestalks suppressed the normal hyperglycemic response, and that stresses like asphyxia produced hyperglycemia only if the sinus gland was present. They were also able to fractionate a crude extract in a Sephadex G-100 column and identify an impure peak that contained the hyperglycemic activity [4]. Kleinholz and Keller [5] were the first to point out that in the crab *Cancer magister* more than one form of hyperglycemic hormone is found. When reversed-phase high performance liquid chromatography (RP-HPLC) was applied to the fractionation of sinus gland extracts [6], it became apparent that in many decapod crustaceans there are at least two isomorphs of the crustacean hyperglycemic hormone (CHH).

The *medulla terminalis*-X organ-sinus gland complex

Decapod crustaceans have a very unique neuroendocrine axis consisting of a cluster of 150–200 peptidergic neurons that form the X-organ in the *medulla terminalis* ganglion of the optic nerve. The axons of these neurons coalesce into the sinus gland nerve that ends in the sinus gland (Fig. 1). The sinus gland is a neurohemal organ consisting of the bulbous axonic terminals loaded with neurosecretory granules that are discharged directly to the hemolymph by a process of exocytosis. In this way, under appropiate physiological or environmental stimuli, different peptide neurohormones are liberated to the general circulation.

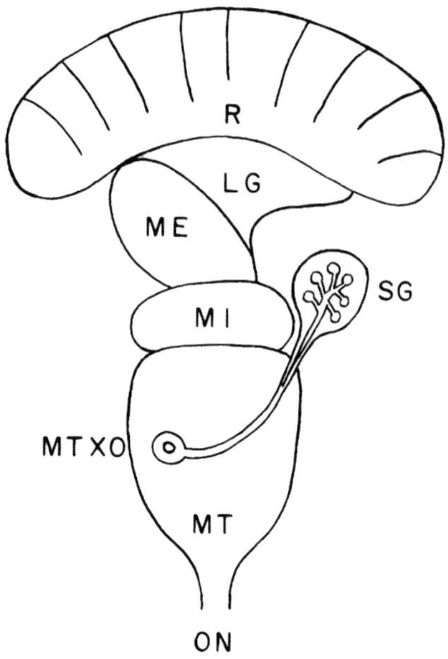

Figure 1. Schematic view of a crustacean eyestalk. ON, optic nerve; MT, *medulla terminalis*; MTXO, *medulla terminalis*-X-organ; MI, *medulla interna*; ME, *medulla externa*; LG, *lamina ganglionaris*; R, retina; SG, sinus gland.

Physiology of CHH

The CHH mediates the increase of hemolymph glucose concentration as a response to physiological or stressful inputs, and also a circadian rhythm of circulating glucose that peaks at night. This circadian rhythm [7, 8] and the response to stress (anoxia, hyperthermia, crowding) are totally abolished in animals devoid of sinus gland.

Upon injection of CHH a decrease of tissue glycogen is observed mainly in abdominal muscle; after ablation of the eyestalk, an increase in tissue glycogen takes place. If the glycogen has been prelabelled with [^{14}C]-glucose, it could be shown that the hepatopancreas releases radioactive glucose under the influence of CHH and the cyclic nucleotides, dibutyryl-cyclic adenosine monophosphate (cAMP) and dibutyryl-cyclic guanosine 5′-phosphate (cGMP). At the same time, the active form of glycogen synthase in abdominal muscle of the eyestalk-ablated crayfish *Orconectes limosus* increases from 30 to 70% without a change in total enzyme. The injection of CHH rapidly inactivates the glycogen synthase and probably activates the phosphorylase in a time course comparable to the hyperglycemia [7, 9–11]. The release of amylase by incubated *O. limosus* hepatopancreas (midgut gland) was stimulated by CHH, dibutyryl-cAMP, dibutyryl-cGMP, Ca^{2+}, Ca^{2+}-ionophore A23187 and the adenylate cyclase activator forskolin [12].

Eyestalk ablation leads to a significant decrease in total lipids in the hemolymph of the crab *Chasmagnathus granulata*, and of free fatty acid, phospholipids and triglyceride levels in the hemolymph of the shore crab, *Carcinus maenas*, and these effects are reversed by the injection of CHH. This hormone increases the release of free fatty acids and phospholipids from *O. limosus* hepatopancreas *in vitro*, suggesting a physiological role for CHH in the control of lipid metabolism in crustaceans [13] (for reviews see refs 23, 36, 37).

Known CHH primary structures

The amino acid sequence of CHH is known in six species of aquatic decapods: *C. maenas* [14]; the lobster *Homarus americanus* [15, 16]; the crayfishes *O. limosus* [17], *Procambarus bouvieri* [18] and *P. clarkii* [19]; the shrimp *Penaeus japonicus* [20]; and one terrestrial isopod, *Armadillidium vulgare* [21]. It is a highly conserved molecule (Fig. 2). It is comprised of 72 amino acid residues in the aquatic crustaceans and of 73 residues in the terrestrial species. Seventeen out of 72 residues (23.6%) are identical in the 11 sequences known to date. If the terrestrial isopod is not included, the identity of the aquatic species raises to 21 residues (29.2%), and if conservative substitutions are allowed, the similarity among all the structures increases to 39 residues (54.2%). Most of the conserved residues occur at the first three-quarters of the molecule, the C-terminus being the most variable region.

The degree of similarity, which ranges from 40.3 to 100%, agrees well with the phylogenetic distance among the organisms [22]. The highest score corresponds to two species of very related American crayfish (*P. bouvieri* and *P. clarkii*), whereas the lowest one results from comparing two species belonging to different infraorders, the terrestrial isopod and the marine shrimp (*A. vulgare* and *Penaeus japonicus*).

The relative degrees of sequence similarity are also in agreement with the interspecies hyperglycemic activities [23]. Thus, the fact that the CHH from *C. maenas* has no effect on one isopod, *Porcellio dilatatus* [24], matches the rather lower score (52.7%) between the CHHs from the shore crab and another isopod, *A. vulgare* (Fig. 2). On the other hand, CHHs from closely related organisms cross-react in bioassays for hyperglycemia. Interspecific activity has been restricted by molecular variations of CHH during crustacean evolution. This is also evident in comparative disk electrophoresis in polyacrylamide gels [22].

The conserved residues of all the CHHs known include six cysteine residues that form three disulfide bonds. The conectivity of the disulfide bonds has been shown to be the same, 7–43 (I–V), 23–29 (II–IV), and 26–52 (III–VI), in four species: *C. maenas*, *Procambarus bouvieri*, *A. vulgare* and *P. clarkii*. They could define a compact and somewhat rigid core (residues

```
               10        20        30        40        50        60        70
Prb   pEVFDQACKGIYDRAIFKKLDRVCEDCYNLYRKPYVATTCRQNCYANSVFRQCLDDLLLIDVVDEYISGVQTV-NH2

HoaA  pEVFDQACKGVYDRNLFKKLDRVCEDCYNLYRKPFVATTCRENCYSNWVFRQCLDDLLLSDVIDEYVSNVQMV-NH2
HoaB  pEVFDQACKGVYDRNLFKKLNRVCEDCYNLYRKPFIVTTCRENCYSNRVFRQCLDDLLMIDVIDEYVSNVQMV-NH2

Cam   pEIYDTSCKGVYDRALFNDLEHVCDDCYNLYRTSYVASACRSNCYSNLVFRQCMDDLLMMDEFDQYARKVQMV-NH2

Pej3   SLFDPACTGIYDRQLLRKLGRLCDDCYNVFREPKVATGCRSNCYHNLIFLDCLEYLIPSHLQEEHMAAMQTV-NH2

Arv    RIFDTSCKGFYDRGLFAQLDRVCEDCYNLYRKPHVAAECRRDCYTTEVFESCLKDLMMHDFINEYKEMALMVS-NH2
          -  - -  --     -   - -- -  -       --  -     -  -                  -
        ===  = ===== ==  = ===========   == ==   == = ==  ==  ===          =  =
```

Figure 2. Amino acid sequence comparison among selected crustacean hyperglycemic hormones. Prb, crayfish, *Procambarus bouvieri*; HoaA and HoaB, lobster, *Homarus americanus*; Cam, shore crab, *Carcinus maenas*; Pej, shrimp, *Penaeus japonicus*; Arv, terrestrial isopod, *Armadillidium vulgare*. Conserved residues are indicated by single dashes at the bottom of the figure. Double dashes highlight both conserved residues and positions where conservative substitutions occur.

7–52), and allow the last quarter of the molecule to adopt a less constrained conformation.

Another conserved molecular feature of the CHHs is the posttranslational modification of their N- and C-termini. An amidated valine occurs at the C-terminus in all aquatic species, whereas an amidated serine is present in the terrestrial one. In addition, the CHHs from five organisms contain a pyroglutamate residue as the N-terminus. The exceptions are the shrimp and the terrestrial isopod.

D-Amino acids in crustacean hyperglycemic hormones

It is a common observation that CHHs appear in pairwise isomorphs (doublet peaks) in all crustaceans where a careful fractionation of sinus gland extracts have been performed by RP-HPLC (*Cardisoma carnifex*, *O. limosus*, *H. americanus*, *P. bouvieri*, *P. clarkii*). These isomorphs are clearly differentiated chromatographically by their degree of hydrophobicity, which allows a certain measure of separation in acetonitrile or isopropanol gradients on octadecyl or phenyl columns. Moreover, in those cases where these isomorphs were isolated and studied by composition, sequence, molecular mass, isoelectric point and circular dichroism, no difference could be found between each pair.

In the lobster *H. americanus*, two variants (CHH-A and CHH-B) of the hyperglycemic hormone were found [25, 26], and their sequence was established by the polymerase chain reaction [27]. Each variant corresponded to a different gene, and each was present as two isomorphs that were purified to homogeneity and found to have identical composition and sequence [15, 28]. Except for the modification of the N- and C-termini (pyroglutamic acid and valinamide, respectively), no other obvious posttranslational modification was apparent.

Peptide mapping experiments uncovered an important difference: when the two isomorphs of CHH-A and CHH-B were digested with endoproteinase Lys-C under reducing conditions, and fractionated in a Nucleosil C18 column, the N-terminal octapeptides were the only ones that eluted with different retention times, but their sequences (pGlu-Val-Phe-Asp-Glu-Ala-Cys-Lys) and their molecular masses (921.4 and 921.0) were the same in both cases.

The only fact that could explain these results was a change of chirality of one or more of the eight residues of the N-terminal octapeptide. This was explored by hydrolyzing the peptides and derivatizing the amino acids with the chiral reagent 1-fluoro-2,4-dinitrophenyl-5-L-alanine. When the derivatized amino acids were separated by RP-HPLC on a Nucleosil C18 column, only phenylalanine showed a difference: it was found to be in the L configuration in the N-terminal peptide from the more polar isomorphs of CHH-A and CHH-B, whereas the D configuration was found in the less polar isomorphs (Fig. 3). This finding was confirmed by chromatographic analysis of the corresponding synthetic peptides containing L- or D-phenylalanine [29].

The two isomorphs of CHH-A were injected into the crayfish *O. limosus*, and the time course of the glycemia measured at intervals during 8 h. It was observed that the injection of 40 pmol of the L-phenylalanine-containing isomorph produced a rapid hyperglycemia with a maximal value attained after 2 h, while the injection of the D-phenylalanine-containing isomorph produced the same maximal value only after 3–4 h and the return to baseline values was slower with the latter hormone.

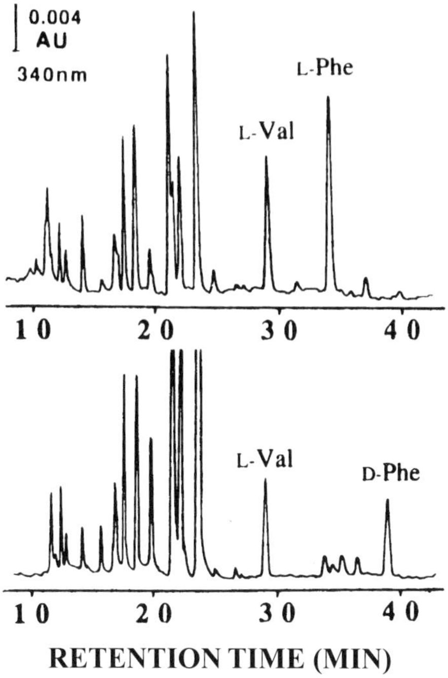

Figure 3. Chiral amino acid analysis of hydrolysates of the N-terminal octapeptides from the first (upper) and the second (lower) isomorphs of CHH-A. For details see text and ref. 29. (Reproduced by permission of the American Society for Biochemistry and Molecular Biology, Inc.).

From the *medulla terminalis*-X organ-sinus gland complex of the crayfish *P. clarkii*, Yasuda et al. [19] have isolated two variants of the crustacean hyperglycemic hormone by RP-HPLC on a TSK gel ODS-120T column. They were designated CHH-I and CHH-II in the order of elution. These isomorphs were digested with endoproteinase Lys-C, and the resulting peptides were chromatographed on the above-mentioned column. Only one peptide (LE1), identified as the N-terminal octapeptide (pGlu-Val-Phe-Asp-Gln-Ala-Cys-Lys), proved to be less polar in the case of CHH-II. Both N-terminal peptides were hydrolyzed, and the resulting amino acids were derivatized with the chiral reagent (+)-1-(9-fluorenyl)ethyl chloroformate (FLEC). The FLEC amino acids were separated on a Wakosil 5C18 column with the result that the third amino acid in the sequence (phenylalanine) in CHH-II belongs to the D-configuration, D-Phe. Both CHH-I and CHH-II increased hemolymph glucose levels when injected into eyestalk-ablated animals, and both had inhibitory activity on ecdysteroidogenesis by Y-organs incubated *in vitro*, but CHH-II was much more potent than CHH-I.

From the sinus glands of the Mexican crayfish *P. bouvieri*, Aguilar et al. [30] have isolated two isomorphs with hyperglycemic activity, named CHH-I and CHH-II in the order of elution from a μ-Bondapak Phenyl RP-HPLC column. These isomorphs had the same composition, sequence, p*I*, molecular mass and circular dichroism. Their only difference was in their respective concentrations (3:1) and lower polarity of CHH-II. When the isomorphs were digested with trypsin and the resulting tryptic peptides were fractionated on an Ultrasphere C18 RP-HPLC column, only one octapeptide (T4) eluted at different times: the one corresponding to CHH-II was less polar (eluted later) than the one corresponding to CHH-I, but their sequences were the same (pGlu-Val-Phe-Asp-Gln-Ala-Cys-Lys), and proved to be the N-termini. Two antibodies were raised in rabbits with synthetic octapeptides (coupled to bovine thyroglobulin) with the above-mentioned sequence, but containing L-phenylalanine or D-phenylalanine as the third residue. The specificity of the antisera was tested by enzyme-linked immunosorbent assay (ELISA) assays and were shown to be highly specific for their respective octapeptides. Both isomorphs were challenged with both antibodies, but CHH-I reacted with the anti-L-octapeptide antiserum, while CHH-II was recognized by the anti-D-octapeptide antiserum (Fig. 4). When the purified CHH-I and CHH-II were injected into eyestalk-ablated animals, the glycemia obtained was the same for both after 1 h. Preliminary observations showed that CHH-II is more effective in the release of amylase by hepatopancreatic cell culture than CHH-I.

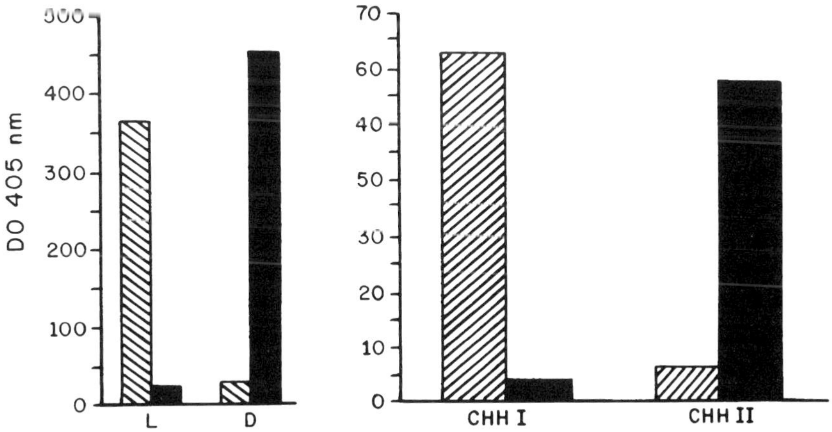

Figure 4. Immunochemical characterization of the CHH isomorphs from the Mexican crayfish, *Procambarus bouvieri*. Left panel: Control assay with 20 ng of synthetic octapeptides pEVFDQACK (L) and pEVdFDQACK (D). Right panel: Assay with 25 ng of CHH-I and CHH-II. The assays were performed with antiserum raised against the L-peptide (striped column) or the D-peptide (solid column). For details see text and ref. 30. (Reproduced by permission of Pergamon Press).

Conclusions

Ribosomal protein biosynthesis implies an absolute chiral requirement for the incorporated amino acids: only the L configuration is accepted by the translational machinery of the cell when following the coded instructions of the messenger RNA.

The posttranslational α-epimerization of a residue in a peptide of ribosomal synthesis has been known since Montecucchi et al. [31] found that dermorphin, an opiatelike heptapeptide from the skin of the South American hylid frog *Phyllomedusa sauvagei* contained a D-alanine in the second position of the peptide (Tyr-D-Ala-Phe-Gly-Tyr-Pro-Ser-NH_2).

Since then, a few other peptides of diverse origins and activities have been found to have D-amino acid residues, mainly in the second position. We can now add four crustacean hyperglycemic hormones that have in common an N-terminal sequence (pGlu-Val-Phe-Asp-Gln-Ala-Cys-Lys) and that are always present as two isomorphs. In the case of the lobster *H. americanus*, there are two hyperglycemic hormones, originated in two genes, each of them present in two isomorphs. In the case of *Procambarus bouvieri* and *P. clarkii*, there is only one gene product, but the hormone is present in two isomorphs. In all these cases, the difference between the isomorphs of the CHH is a posttranslational modification consisting of an α-epimerization of the third residue, phenylalanine. This difference could explain the difference in hydrophobicity of the intact hormones and also the difference in mobility of the N-terminal octapeptides. We have evidence that in *P. bouvieri* the molt-inhibiting hormone (MIH) is also present as two isomorphs differentiated by a D-phenylalanine in the third position (A. Huberman, unpublished observations). It is possible that this posttranslational modification is more widespread in crustaceans than hitherto thought. In practically all peptides known to have a D-amino acid residue, this is situated close to the N-terminus, either in the second or third position. There is only one known case of a D-amino acid residue close to the C-terminus of a peptide: ω-Aga IVB from the venom of the spider *Agelenopsis aperta*, which contains a D-serine in position 46 in a total of 48 residues, that is, in the third position from the C-terminus.

The physiological meaning of this type of isomorphism is not readily apparent. In the case of the lobster *H. americanus*, it has been shown [29] that the D-amino acid residue-containing isomorph produces a slower and longer-lasting hyperglycemia, but the maximal level of hemolymph glucose is the same with both isomorphs, and this could be consistent with a slower inactivation or degradation of the modified CHH. Yasuda et al. [19] have seen a more intense inhibition of ecdysteroid biosynthesis by Y-glands incubated *in vitro* by the D-amino acid residue-containing isomorph of the CHH of *P. clarkii*. In the case of the CHHs of *P. bouvieri*, Aguilar et al. [30] did not see a different effect of both isomorphs on the glycemia at short times (1 h), but there was an indication that the D-amino acid residue-containing isomorph (CHH-II) induced a stronger release of amylase from a culture of hepatopancreas cells (A. Huberman, unpublished observations).

Recently, [32, 33] an enzyme (a peptidyl-aminoacyl-L-D-isomerase) was isolated from the venom of the funnel web spider *Agelenopsis aperta* that converts, reversibly, an L-serine to a D-serine in the ω-Aga IVB peptide.

It is important to point out that only one residue per peptide is modified. Other amino acids of the same kind along the peptide remain as L-amino acid residues. This could mean that there is a sequence or tridimensional structure that is recognized by the enzyme. As to the mechanism of action of this enzyme, it has been proposed that it acts by a process of dehydrogenation and stereospecific hydrogenation of the α-carbon, inverting its configuration (for reviews, see refs 34, 35).

The enzymatic posttranslational modification that converts an L-amino acid residue, already incorporated into a peptide, to the corresponding D-isomer in many different organisms (invertebrate and vertebrate) must have an incompletely understood importance that should be explored at the organ and receptor levels. In the case of CHH, it is possible that the two isomorphs have a differential action on specific tissues and organs.

Acknowledgements
The research described here was made possible by grants from the Consejo Nacional de Ciencia y Tecnología (CONACYT), and the Mexico-United States Foundation for Science to A. Huberman.

References

1. Abramowitz, A.A., Hisaw, F.L. and Papandrea, D.N. (1944) The occurrence of a diabetogenic factor in the eyestalks of crustaceans. *Biol Bull* 86:1–5.
2. Kleinholz, L.H. and Little, B.C. (1949) Studies in the regulation of blood-sugar concentration in crustaceans. I. Normal values and experimental hyperglycemia in *Libinia emarginata. Biol Bull* 96:218–227.
3. Kleinholz, L.H., Havel, V.J. and Reichart, R. (1950) Studies in the regulation of blood-sugar concentration in crustaceans. II. Experimental hyperglycemia and the regulatory mechanisms. *Biol Bull* 99:454–468.
4. Kleinholz, L.H., Kimball, F. and McGarvey, M. (1967) Initial characterization and separation of hyperglycemic (diabetogenic) hormone from the crustacean eyestalk. *Gen Comp Endocrinol* 8:75–81.
5. Kleinholz, L.H. and Keller, R. (1973) Comparative studies in crustacean neurosecretory hyperglycemic activities. I. The initial survey. *Gen Comp Endocrinol* 21:551–564.
6. Keller, R. and Kegel, G. (1984) Studies on crustacean neuropeptides by use of high performance liquid chromatography. In: J.A. Hoffman and M. Porchet (eds): *Biosynthesis, metabolism and mode of action of invertebrate hormones*. Springer-Verlag, Heidelberg, 145–154.
7. Keller, R. (1965) Uber eine hormonale Kontrolle des Polysaccharidstoffwechsels beim Flusskrebs *Cambarus affinis* Say. *Zeit Vergl Physiol* 51:49–59.
8. Hamann, A. (1974) Die neuroendokrine Steuerung tagesrhythmischer Blutzuckerschwankungen durch die Sinusdrüse beim Flusskrebs. *J Comp Physiol* 89:197–214.
9. Sedlmeier, D. (1987) The role of hepatopancreatic glycogen in the action of the crustacean hyperglycemic hormone (CHH). *Comp Biochem Physiol* 87A:423–425.
10. Sedlmeier, D. and Keller, R. (1981) The mode of action of the crustacean neurosecretory hyperglycemic hormone. I. Involvement of cyclic nucleotides. *Gen Comp Endocrinol* 45:82–90.

11. Sedlemeier, D. (1982) The mode of action of the crustacean neurosecretory hyperglycemic hormone (CHH). II. Involvement of glycogen synthase. *Gen Comp Endocrinol* 47:426–432.
12. Sedlemeier, D (1988) The crustacean hyperglycemic hormone (CHH) releases amylase from the crayfish midgut gland. *Regul Pept* 20:91–98.
13. Santos, E. A., Nery, L. E. M., Keller, R. and Goncalves, A. A. (1997) Evidence for the involvement of the crustacean hyperglycemic hormone in the regulation of lipid metabolism. *Physiol Zool* 70:415–420.
14. Kegel, G., Reichwein, B., Weese, S., Gaus, G., Peter-Katalinic, J. and Keller, R. (1989) Amino acid sequence of the crustacean hyperglycemic hormone (CHH) from the shore crab, *Carcinus maenas*. *FEBS Lett* 255:10–14.
15. Tensen, C. P., Janssen, K. P. C., Soyez, D. and Van Herp, F. (1991) Comparative characterization of hyperglycemic neuropeptides from the lobster *Homarus americanus*. *Peptides* 12: 241–249.
16. Chang, E.S., Prestwich, G.D. and Bruce, M.J. (1990) Amino acid sequence of a peptide with both molt-inhibiting and hyperglycemic activities in the lobster, *Homarus americanus*. *Biochem Biophys Res Comm* 171:818–826.
17. Kegel, G., Reichwein, B., Tensen, C. P. and Keller, R. (1991) Amino acid sequence of crustacean hyperglycemic hormone (CHH) from the crayfish, *Orconectes limosus*: emergence of a novel neuropeptide family. *Peptides* 12:909–913.
18. Huberman, A., Aguilar, M.B., Brew, K., Shabanowitz, J. and Hunt, D.F. (1993) Primary structure of the major isomorph of the crustacean hyperglycemic hormone (CHH-I) from the sinus gland of the Mexican crayfish *Procambarus bouvieri* (Ortmann): interspecies comparison. *Peptides* 14:7–16.
19. Yasuda, A., Yasuda, Y., Fujita, T. and Naya, Y. (1994) Characterization of crustacean hyperglycemic hormone from the crayfish (*Procambarus clarkii*): multiplicity of molecular forms by stereoinversion and diverse functions. *Gen Comp Endocrinol* 5:387–398.
20. Yang, W.-J., Aida, K. and Nagasawa, H. (1995) Amino acid sequences of hyperglycaemic hormone and its related peptides from the Kuruma prawn, *Penaeus japonicus*. *Aquaculture* 135:205–212.
21. Martin, G., Sorokine, O. and Van Dorsselaer, A. (1993) Isolation and molecular characterization of a hyperglycemic neuropeptide from the sinus gland of the terrestrial isopod *Armadillidium vulgare* (Crustacea). *Eur J Biochem* 211:601–607.
22. Keller, R., Jaros, P. P. and Kegel, G. (1985) Crustacean hyperglycemic neuropeptides. *Amer Zool* 25:207–221.
23. Keller, R. (1992) Crustacean neuropeptides: structures, functions and comparative aspects. *Experientia* 48:439–448.
24. Martin, G., Keller, R., Kegel, G., Beese, G. and Jaros, P.P. (1984) The hyperglycemic neuropeptide of the terrestrial isopod, *Porcellio dilatatus*. I. Isolation and characterization. *Gen Comp Endocrinol* 55:208–216.
25. Soyez, D., Van Deijnen, J.E. and Martin, M. (1987) Isolation and characterization of a vitellogenesis inhibiting factor from the sinus gland of the lobster, *Homarus americanus*. *J Exp Zool* 244:479–484.
26. Tensen, C.P., Janssen, K.P.C. and Van Herp, F. (1989) Isolation, characterization and physiological specificity of the crustacean hyperglycemic hormone from the sinus gland of the lobster, *Homarus americanus* (Milne-Edwards). *Invert Reprod Dev* 16: 155–164
27. Tensen, C.P., De Kleijn, D.P.V. and Van Herp, F. (1991) Cloning and sequence analysis of cDNA encoding two crustacean hyperglycemic hormones from the lobster, *Homarus americanus*. *Eur J Biochem* 200:103–106.
28. Soyez, D., Noël, P.Y., Van Deijnen, J.E., Martin, M., Morel, A. and Payen, G.G. (1990) Neuropeptides from the sinus gland of the lobster, *Homarus americanus*: characterization of hyperglycemic peptides. *Gen Comp Endocrinol* 79:261–274.
29. Soyez, D., Van Herp, F., Rossier, J., Le Caer, J.-P., Tensen, C.P. and Lafont, R. (1994) Evidence for a conformational polymorphism of invertebrate neurohormones. D-amino acid residue in crustacean hyperglycemic peptides. *J Biol Chem* 269:18295–18298.
30. Aguilar, M.B., Soyez, D., Falchetto, R., Arnott, D., Shabanowitz, J., Hunt, D.F. and Huberman, A. (1995) Amino acid sequence of the minor isomorph of the crustacean hyperglycemic hormone (CHH-II) of the Mexican crayfish *Procambarus bouvieri* (Ortmann): presence of a D-amino acid. *Peptides* 16:1375–1383.

31. Montecucchi, P.C., De Castiglione, R., Piani, S., Gozzini, L. and Erspamer, V. (1981) Amino acid composition and sequence of dermorphin, a novel opiate-like peptide from the skin of *Phyllomedusa sauvagei*. *Int J Peptide Prot Res* 17:275–283.
32. Shikata, Y., Watanabe, T., Teramoto, T., Inoue, A., Kawakami, Y., Nishizawa, Y., Katayama, K. and Kuwada, M (1995) Isolation and characterization of a peptide isomerase from funnel web spider venom. *J Biol Chem* 270:16719–16725.
33. Heck, S.D., Faraci, W.S., Kelbaugh, P.R., Saccomano, N.A., Thadeio, P.F. and Volkmann, R.A. (1996) Posttranslational amino acid epimerization: enzyme catalyzed isomerization of amino acids in peptide chains. *Proc Natl Acad Sci USA* 93:4036–4039.
34. Mor, A., Amiche, A. and Nicolas, P. (1992) Enter a new post-translational modification: D-amino acids in gene-encoded peptides. *Trends Biochem Sci* 17:481–485.
35. Kreil, G. (1997) D-amino acids in animal peptides. *Ann Rev Biochem* 66:337–345.
36. Santos, E.A. and Keller, R. (1993) Crustacean hyperglycemic hormone (CHH) and the regulation of carbohydrate metabolism: current perspectives. *Comp Biochem Physiol* 106A:405–411.
37. Fingerman, M. (1997) Crustacean endocrinology: a retrospective, prospective and introspective analysis. *Physiol Zool* 70:257–269.

Biosynthesis:
A new family of isomerases

D-Amino Acids in Sequences of Secreted Peptides of Multicellular Organisms
ed. by P. Jollès

Biosynthesis of D-amino acid-containing peptides: Exploring the role of peptide isomerases

R. A. Volkmann and S. D. Heck

Pfizer Central Research, Eastern Point Road, Groton, CT 06340, USA

Summary. The discovery of D-amino acid residues in a growing number of gene-encoded peptides suggests that such biochemical modifications are more common than initially thought. In fact, the extent to which D-amino acids are incorporated into peptides by multicellular organisms probably has not been fully realized, since routine Edman sequencing does not provide the absolute stereochemistry of amino acid residues. Unless both the D and L isomers of a particular peptide sequence are isolated, D-amino acid-containing peptides are often identified only after synthesis of naturally-occurring peptide fails to yield the desired activity.

To date, D-amino acid residues (e.g., alanine, methionine, leucine, isoleucine, phenyl alanine, asparagine, tryptophan and serine) have been identified in peptides from a variety of species, including frogs, snails, clams, lobsters and spiders. While most have a single D-amino acid residue located near their N-termini, an exception is found with ω-Aga IVB. The examples highlighted in this chapter are the result of a unique strategy of multicellular organisms to circumvent stereochemical limitations imposed by the genetic code in an effort to increase molecular diversity. The presence of D-amino acids permits the generation of novel tertiary structure that could not be accessed from L-amino acids alone. Moreover, advantages of increased potency and protease stability are often observed.

Our understanding of the biosynthesis of these D-amino acid-containing peptides is still in its infancy. Nevertheless, the discovery of a novel peptide isomerase from the venom of the *Agelenopsis aperta* spider provides some important clues to explain the incorporation of single D-amino acid residues within a peptide chain. Given its high homology with other serine proteases, the isomerase may represent an opportune mutation in response to evolutionary pressures. Yet, is the isomerase a unique exception or simply the first in a class of enzymes of varying substrate specificity capable of synthesizing D-amino acid-containing peptides? To be sure, much more remains to be explored about the precise timing and mechanism of the isomerization process, in addition to obtaining further structural data on the enzyme itself. Therein lies the continuation of this fascinating story in enzyme biochemistry.

Introduction

Nature has provided multicellular organisms with a remarkable strategy for generating peptide diversity by utilizing a pool of 21 amino acids and genetic information encoded in the nucleotide sequence of messenger RNA. Biosynthesis, however, does not end with transcription and translation; nearly all proteins are posttranslationally modified during or after biosynthesis. Acylation, glycosylation, deformylation, amidation, phosphorylation, sulfation, methylation, hydroxylation and iodination reactions are among a number of biochemical pathways available to eukaryotes for introducing over 200 amino acid variants in proteins. These modifications, by and large, are essential for biological specificity and function.

The discovery of D-amino acid residues in naturally occurring peptides has suggested yet another biochemical pathway for creating peptide diversity. D-Amino acid residues (D-Ala and D-Glu) comprise the cell wall of bacteria, are biosynthetic precursors (D-Val) to penicillins and are present in a number of larger antibiotics that are synthesized in microorganisms. Biosynthetic studies of gramicidines, tyrocidines and other bacterial-derived peptides have shown that these structures are generated by nonribosomal pathways in a stepwise fashion through a series of discrete biochemical transformations [1, 2]. The mechanism for racemization primarily involves the intermediacy and racemization of an amino acid thioester (presumably derived from pantheine thiols) and its subsequent incorporation into peptides via a multienzymic process.

In recent years, additional D-amino acid-containing peptides have been characterized. Initially, their emergence was regarded as having nonribosomal origins. Those emanating from vertebrates and invertebrates were considered artifacts or the result of diet or long-term amino acid (Asp/Ser) racemization in aged proteins. However, the isolation of a family of complex peptide polycyclic antibiotics named lantibiotics (nisin, subtilin, epidermin) from Gram-positive bacteria dispelled the notion that unicellular organisms were the sole source of peptides containing D-amino acids. Biosynthetic studies indicated that these dehydro- and D-amino acid-containing peptides are products of posttranslational processing from ribosomally synthesized (gene-encoded) precursors. Subsequent studies revealed the intermediacy of L-serine and L-threonine-derived dehydroamino acids (via dehydration) in the generation of these D-amino acid residues. The spectacular series of events that have conspired to provide the lantibiotics are discussed in detail elsewhere [3, 4].

A growing list of D-amino acid-containing peptides has emerged from a variety of eukaryotic sources over the last two decades. In general, these peptides have a single D-amino acid residue located near their N-termini. This residue is a key determinant in the overall structure of the molecule and is a prerequisite for biological activity. D-Amino acid residues have now been found in peptides isolated from such diverse sources as frogs, snails, clams, lobsters and spiders. Their presence has revealed a fascinating strategy utilized by these species for generating biological diversity and, in the process, circumventing the restrictions imposed by gene-encoded peptide syntheses which utilize the L-amino acid pool.

Excellent reviews describing aspects of the posttranslational isomerization of amino acid residues in gene-encoded peptides have recently appeared [5–7]. Current advances, particularly those involving the isolation and characterization of a peptide isomerase from the funnel web spider, *Agelenopsis aperta*, have provided exciting new clues to explain how multicellular organisms are able to incorporate D-amino acids into peptide structures. In the course of this chapter, we intend to highlight what is currently known about the D-amino acid-containing peptides that have been isolated from vertebrates and invertebrates and include thoughts on how multicellular organisms are able to alter the absolute configuration of amino acid residues in a peptide sequence.

Dermorphins, deltorphins and related amphibian peptides

Examination of skin secretions from various amphibian species has led to the discovery of a host of novel, biologically active peptides [8, 9]. Many of these structures have their counterparts in mammalian tissues, interacting primarily with receptors of the central and peripheral nervous system and gastrointestinal tract. Cutaneous tissue from neotropical hylid frogs belonging to the genus *Phyllomedusinae* contains an especially rich diversity of compounds from a number of peptide families [10]. Examples include the caeruleins, tachykinins, bradykinins, bombesins, tryptophyllins and sauvagine. These and other hormonelike structures have been extensively studied by Erspamer and colleagues. Their efforts represent the beginning of a fascinating story in comparative biochemistry-namely the discovery and biological characterization of a series of vertebrate peptides containing D-amino acids.

In 1981, Erspamer and co-workers isolated dermorphin, a novel heptapeptide, from skin extracts of the Argentinian tree frog *Phyllomedusa sauvagei* [11]. This structure was the first in a new class of peptides displaying intense, long-lasting opioid analgesia. Unlike endogenous opioid ligands found in vertebrate brain, dermorphin is highly selective for μ opiate receptors, showing little affinity for δ or κ receptors [12]. Intracerebroventricular injections of dermorphin in rats, mice and rabbits found it to be up to 1000-fold more potent than morphine and over 10,000-fold more potent than the enkephalins on a molar basis [10].

Initial attempts to prepare dermorphin were unsuccessful, as synthetic samples having the sequence Tyr-Ala-Phe-Gly-Tyr-Pro-Ser-NH_2 were completely inactive [13]. A closer examination of the structure revealed a D-alanine residue in the second position from the N-terminus of natural peptide [11]. A related peptide, having a hydroxyproline in position 6 in addition to a D-alanine in position 2, was also isolated from the same source and found to be equipotent to dermorphin [14].

The discovery of dermorphin represented the first known example of the incorporation of a D-amino acid residue in a eukaryotic peptide. To be sure, the presence of D-alanine raised many questions concerning its biosynthesis. Was it the product of a thiotemplate multienzymic mechanism as employed in the nonribosomal synthesis of peptide antibiotics [15]? Alternatively, could it be derived from a posttranslational modification of a ribosomally synthesized biosynthetic precursor comparable to the lantibiotics [16]?

These questions were answered with the construction of complementary DNA (cDNA) libraries from *P. sauvagei* frog skin messenger RNA (mRNA) [17]. Amplified clones were screened with a mixture of oligonucleotides containing the complementary codons for the Phe-Gly-Tyr-Pro-Ser segment of dermorphin. The pro-dermorphin sequence was provided with the isolation of several clones containing inserts coding for fragments of dermorphin precursors. Typical of a pro-hormone, a Lys-Arg processing signal is followed by a 35-amino acid repeat coding for a single dermorphin sequence. A Gly-Glu-Ala-Lys-Lys cleavage signal exposes a C-terminal glycine, giving rise to the carboxyamidation of the heptapeptide [18]. However, unlike the case of the lantibiotics, the cloned cDNA encoding the pro-dermorphin precursor contains the usual L-Ala codon GCG in the D-Ala position [17]. This observation provided early evidence to suggest a posttranslational isomerization of an amino acid from the L to D configuration.

Sequencing of additional cloned cDNAs detected a dermorphin precursor in which four of five homologous repeats code for dermorphin, while the other (first) codes for a different heptapeptide, having the predicted sequence Tyr-Met-Phe-His-Leu-Met-Asp-NH_2 (Fig. 1) [17]. Inspection of *P. sauvagei* skin extracts, in fact, revealed the presence of small quantities of this heptapeptide. As in the case of dermorphin, the second amino acid (methionine) from the N-terminus of the molecule exists in the D configuration. This structure, named deltorphin, was the first in another class of

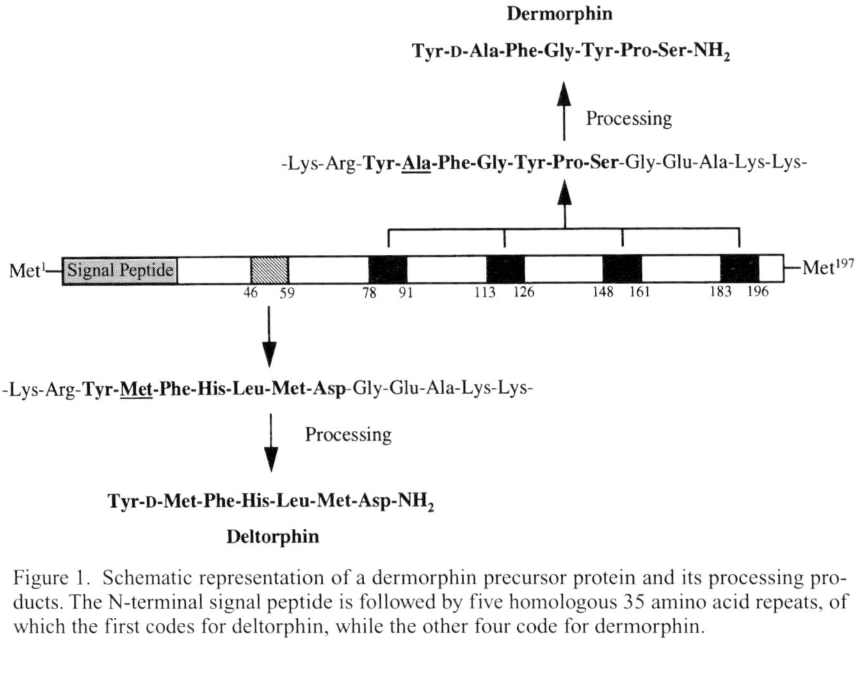

Figure 1. Schematic representation of a dermorphin precursor protein and its processing products. The N-terminal signal peptide is followed by five homologous 35 amino acid repeats, of which the first codes for deltorphin, while the other four code for dermorphin.

Phyllomedusa peptides shown to be highly potent and selective for δ opiate receptors [19–22]. Binding studies with several enkephalin analogs found the affinity and potency of deltorphin to be among the highest of all the δ-selective ligands known to date [23].

Since these initial discoveries, a number of endogenous opioid peptides belonging to the dermophin and deltorphin families have been identified (Tab. 1) [9, 24]. Each of these amphibian skin peptides has an N-terminal sequence Tyr-Xaa-Phe, in which Xaa corresponds to D-Ala or D-Met. A related 17-amino acid peptide, [D-Leu-2]deltorphin, has been isolated from the skin of a Brazilian tree frog, *Phyllomedus burmeisteri* [25]. While the peptide displays poor affinity for δ-binding sites, a synthetic analog comprised of residues 1–10 was found to be a potent and selective δ-opioid agonist. In each of these examples, the molecules are derived from precursor polypeptides coded for by the corresponding L-amino acid codons, providing mounting evidence to suggest a posttranslational stereoinversion of the L-amino acid residue to its D isomer.

While the precise timing and mechanism of this transformation remain a mystery, isolation and characterization of a dermorphin precursor have provided some clues [26]. Size exclusion chromatography of *P. sauvagei* skin extracts produced fractions having immunoreactivity towards antibodies raised to a 13-amino acid portion of the predicted pro-dermorphin sequence. Purification and proteolytic digestion of the polypeptide precursor generated a peptide containing a D-alanine residue in the same position as in mature peptide. This result is intriguing and may suggest that stereoinversion occurs early in the biosynthesis of dermorphin at the precursor level. Interestingly, the same study also identified [L-Met-2]deltorphin as an endogenous constituent of *Phyllomedusa* skin. Given that the peptide lacks opioid activity, its presence in approximately 100-fold excess over the D isomer may be an indication of the low efficiency of the epimerization process.

Table 1. Amino acid sequences of naturally occurring D-amino acid-containing peptides isolated from amphibian skin extracts

Dermorphin	Tyr-**D-Ala**-Phe-Gly-Tyr-Pro-Ser-NH_2
Dermorphin-OH	Tyr-**D-Ala**-Phe-Gly-Tyr-Pro-Ser-OH
[Hyp-6]Dermorphin	Tyr-**D-Ala**-Phe-Gly-Tyr-Hyp-Ser-NH_2
[Lys-7]Dermorphin	Tyr-**D-Ala**-Phe-Gly-Tyr-Pro-Lys-OH
[Trp-4,Asn-7]Dermorphin	Tyr-**D-Ala**-Phe-Trp-Tyr-Pro-Asn-OH
[Met-2]Deltorphin	Tyr-**D-Met**-Phe-His-Leu-Met-Asp-NH_2
[Met(O)-2]Deltorphin	Tyr-**D-Met(O)**-Phe-His-Leu-Met-Asp-NH_2
[Ala-2]Deltorphin I	Tyr-**D-Ala**-Phe-Asp-Val-Val-Gly-NH_2
[Ala-2]Deltorphin II	Tyr-**D-Ala**-Phe-Glu-Val-Val-Gly-NH_2
[Leu-2]Deltorphin	Tyr-**D-Leu**-Phe-Ala-Asp-Val-Ala-Ser-Thr-Ile-Gly-Asp-Phe-Phe-His-Ser-Ile-NH_2
Bombinin H3 (*Bombina variegata*)	Ile,Leu-**D-alle**-Gly-Pro-Val-Leu-Gly-Met–^{20}Ile-NH_2
Bombinin H4	Ile,Leu-**D-alle**-Gly-Pro-Val-Leu-Gly-Leu–^{20}Ile-NH_2
Bombinin H5	Ile-**D-alle**-Gly-Pro-Val-Leu-Gly-Leu–^{20}Ile-NH_2

In addition to the central nervous system (CNS)-mediated effects of the opioid peptides, another class of amphibian peptides displaying anti-microbial and hemolytic activity has been characterized. Examination of skin secretions from the European frog *Bombina variegata* identified a 24-amino acid peptide, named bombinin, and several related 27-amino acid peptides with antibacterial activity [27]. A cDNA library prepared from the skin of *B. variegata* produced a number of clones coding for precursors of the bombinin family of peptides and led to the identification of an additional series of 20-amino acid hydrophobic and hemolytic peptides, termed bombinins-H [28]. In three of these structures (Tab. 1), the presence of D-alloisoleucine was detected in the second position from the N-terminus.

D-Amino acid-containing peptides from molluscs and crustaceans

Molluscs have proven to be a rich source of neuropeptides and, in addition, have provided another source of D-amino acid-containing peptides. In 1989, Nomoto, Kamatani and co-workers reported the isolation of an endogenous neuroexcitatory tetrapeptide from the subesophageal and cerebral ganglia of the African giant snail *Achatina fulica* F'erussac [29, 30]. This peptide, achatin-1 (Gly-D-Phe-Ala-Asp-OH), contains a D-phenyl alanine residue which is essential for the peptide's marked effects on over half of the 23 types of essential *Achatina* neurons assayed. The D-amino acid chirality is responsible for the molecule's β-turn 15-membered ringlike structure stabilized by two hydrogen bonds involving NH(Gly^1) and C^{γ}=O^{δ}(Asp-4) and NH(Asp-4) and C^{γ}=O^{δ}(Asp-4), as revealed by the single-crystal X-ray structure shown in Figure 2 [31, 32]. Endogenous achatin-2 (Gly-Phe-Ala-Asp-OH), the all-L-amino acid version of achatin-1, has an extended

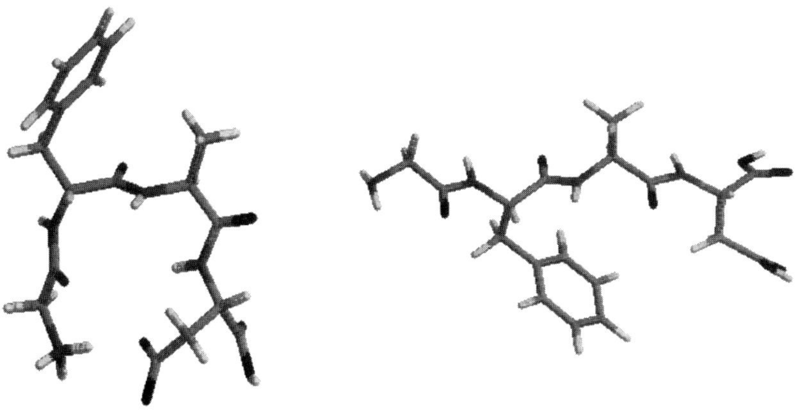

Achatin-1 **Achatin-2**

Figure 2. Comparison of the crystal structures of achatin-1 and achatin-2.

β-pleated structure stabilized by five intermolecular hydrogen bonds and is devoid of achatin-1's neuroexcitatory activity [33]. The precursor protein of achatin-1 is not known.

A second D-amino acid-containing peptide, fulicin (Phe-D-Asn-Glu-Phe-Val-NH_2), was later isolated by Ohta and colleagues from the snail's ganglia [34]. Once again, the D-configuration of the asparagine moiety adjacent to the N-terminus is crucial for biological activity (contraction of the penis retractor muscle). Interestingly, replacement of this residue with other D-amino acids results in little loss of activity, suggesting that an important role of the amino acid's configuration might be to promote receptor binding by stabilizing the peptide's bioactive conformation [35].

A mixture of oligonucleotides coding for fulicin (Phe-Asn-Glu-Phe-Val-Gly) has been used to screen an *Achatina fulica* cDNA library [36]. Restriction mapping and partial sequencing of positive clones showed positive overlap, and a detailed sequence analysis was obtained for two clones, the longest having 1995 nucleotides and encoding a fulicin precursor protein. Unlike the dermorphin precursor protein and other precursor proteins for small bioactive peptides that contain several copies of the same peptide, the fulicin precursor is unique in that it contains one copy of fulicin and one copy of at least nine other putative α-amidated neuropeptides identified as FGRPs (fulicin gene-related peptides) shown in Figure 3. Fulicin is located at the N-terminus, while the FGRPs are located in the C-terminal portion of the precursor protein. The D-asparagine residue present in fulicin is coded for by the usual L-Asn codon (AAT). Each of the N-termini of the FGRPs contains either basic Lys-Arg or Lys-Lys residues, while the C-termini contain Gly-Lys-Arg or Gly-Lys-Lys residues, which serve as consensus sequences for protein processing and subsequent C-terminal amidation. RNAs for the ganglia, atria and ventricles have been prepared, and northern blot analysis provides a band for each, comparable in size to the cDNA clone, suggesting that fulicin as well as the FGRPs are biosynthesized both in neurons and the heart.

A chromatographic comparison of synthetic and native FGRPs isolated from the snail atria resulted in the identification of a third D-amino acid-containing peptide (Tyr-D-Ala-Glu-Phe-Leu-NH_2), fulyal, which also potentiates tetanic contractions of the snail penis retractor muscle at concentrations as low as 10^{-12} M, comparable to fulicin [37]. The primary sequence of fulyal can be predicted from the nucleotide sequence of the cDNA sequence encoding the fulicin precursor protein. Again, D-alanine is coded for by an L-Ala codon (GCA).

Another D-amino acid-containing neuroexcitatory decapeptide was isolated from the anterior byssus retractor muscle of the bivalve mollusc, *Mytilus edulis*, having the primary sequence Ala-Leu-Ala-Gly-Asp-His-Phe-Phe-Arg-Phe-NH_2 [38]. Synthetic efforts to generate this peptide revealed that the second residue of the sequence is not L-Leu but, rather, D-Leu. The all-L-peptide has not been isolated from this clam.

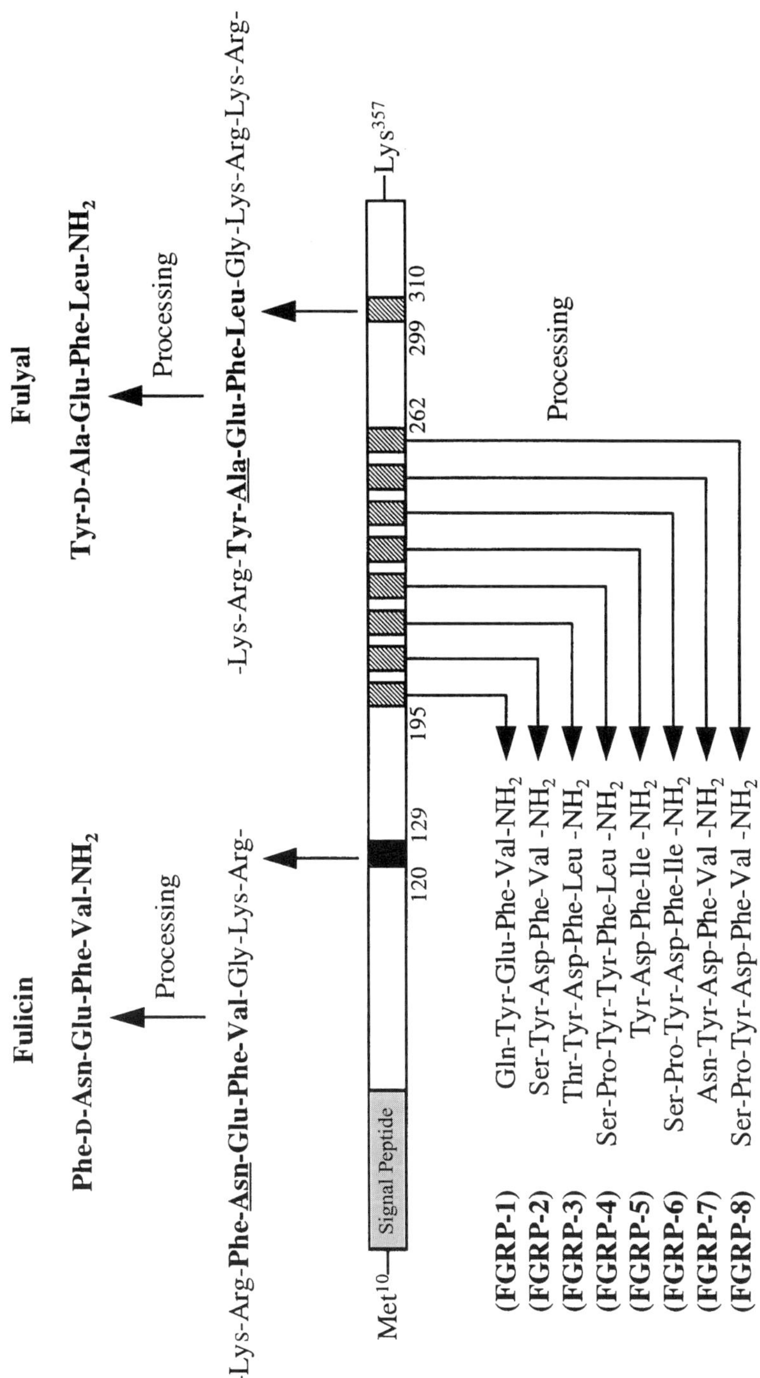

Figure 3. Schematic representation of a fulicin precursor protein and its processing products.

Collectively, the endogenous D-amino acid-containing mollusc and opioid peptides are relatively small (≤ 10 amino acids) with the D-amino acid residue residing adjacent to the N-terminus. In the majority of cases, the D-amino acids confer important biological activity, presumably by stabilizing specific molecular conformations that are important for biological activity and, no doubt in the process, by protecting these small peptides from aminopeptidases. Inspection of the fulicin and opioid peptide cDNA libraries reveals that the neutral Met, Asn and Ala residues affected are all encoded by the usual L-amino acid codons, suggesting a posttranslational mechanism for the conversion of L- to D-amino acids. Unfortunately, the diversity of D-amino acid-containing structures has yielded few mechanistic clues to explain how individual amino acids in the peptide chains are targeted and how the chirality of these residues is altered. With the exception of the neuroexcitatory *Mytilus edulis* decapeptide (Ala-D-Leu-Ala-Gly-Asp-His-Phe-Phe-Arg-Phe-NH_2), the endogenous D-amino acid-containing peptides from molluscs and crustaceans (Tab. 2) are accompanied by the corresponding L-amino acid-containing peptides. The relative amounts of each peptide vary. It has been suggested that the Gly-Glu-Ala-Lys-Lys sequence of residues found at the C-terminus of the dermorphins/deltorphins is a prerequisite for the isomerization process. The same terminus, however, is not present in achatin-1, fulicin or fulyal.

The discovery of 72-residue D-amino acid-containing crustacean hyperglycemic hormones (CHHs) dispelled the notion that epimerization always occurs at the second residue from the N-terminus. One such peptide, named Prc-CHH-II, was isolated from a crayfish (*Procambarus clarkii*), together with the all-L-amino acid peptide [39]. In Prc-CHH-II, a phenyl alanine residue three amino acids from the N-terminus is isomerized. Identical hormones or molecular variants of these neurohormones have been reported in other crayfish (*Orconectes limosus* [40]; *P. bouvieri* [41]) and lobsters (*Homarus americanus*) [42]. These neuropeptides regulate glycemia through glycogen mobilization. The D-amino acid residue plays an important role in the pharmacokinetics of the molecule. When injected into

Table 2. Amino acid sequences of naturally occurring D-amino acid-containing peptides isolated from molluscs and crustaceans

Peptide	Sequence
Achatin-1 (African giant snail, *A. fulica*)	Gly-**D-Phe**-Ala-Asp-OH
Fulician (African giant snail, *A. fulica*)	Phe-**D-Asn**-Glu-Phe-Val-NH_2
Fulyal (African giant snail, *A. fulica*)	Tyr-**D-Ala**-Glu-Phe-Leu-NH_2
Mytilus-FFRFamide (bivale, *M. edulis*)	Ala-**D-Leu**-Ala-Gly-Asp-His-Phe-Phe-Arg-Phe-NH_2
Prc-CHH-II (crayfis, *P. clarkii*)	pGlu-Val-**D-Phe**-Asp-Gln-Ala-Cys-Lys–^{72}Val-NH_2
D-Phe^3 cHH$_A$ (lobster, *H. americanus*)	pGlu-Val-**D-Phe**-Asp-Gln-Ala-Cys-Lys–^{72}Val-NH_2
D-Phe^3 cHH$_B$ (lobster, *H. americanus*)	pGlu-Val-**D-Phe**-Asp-Gln-Ala-Cys-Lys–^{72}Val-NH_2
Contryphan (fish-hunting snail, *C. radiatus*)	Gly-Cys-Hpy-**D-Trp**-Glu-Pro-Trp-Cys-NH_2
Des-[Gly^1]contryphan (fish-hunting snail, *C. radiatus*)	Cys-Hyp-**D-Trp**-Glu-Pro-Trp-Cys-NH_2

crayfish, the D-amino acid isoforms of cHHA (lobster) increase the length of time for maximal response over the L-amino acid isoform [43].

Fish-hunting cone snails contain a multiplicity of relatively small (< 30 amino acid), disulfide-rich peptide toxins that block neuromuscular transmission, leading to rapid immobilization of fish. Olivera, Gray, Cruz and co-workers have isolated and characterized a growing number of interesting toxins from these molluscs, which have demonstrated remarkable specificity for target channels and elicit unique biological effects when administered to animals. They have recently isolated two D-Trp-containing peptides, des-[Gly^{1}]contryphan and contryphan (Gly-Cys-Hyp-D-Trp-Glu-Pro-Trp-Cys-NH_2) from *Conus radiatus* [44]. D-Trp is present in the third and fourth residues of these small disulfide-containing toxins. Analysis of a cDNA clone from a *C. radiatus* venom duct reveals that the D-Trp residues of these peptides are coded for by the standard L-Trp codon. While pairs of Trp and Pro residues are present in both of these toxins, only those near the N-terminus are posttranslationally modified (Pro/Hyp and L/D-Trp). Contryphans are distributed widely in *Conus* and homologous peptides, suggesting that contryphan and des-[Gly^{1}]contryphan might well be the first of a number of D-amino acid-containing peptides to be isolated from *Conus* venoms.

Discovery of the first peptide isomerase

Venomous animals have developed a variety of ingenious chemical strategies for predation. Given that the primary function of venom is the rapid paralysis of prey, the toxins they contain interact directly with many components necessary for synaptic transmission at the neuromuscular junction. The venom of the American funnel-web spider *Agelenopsis aperta* contains a collection of peptide and polyamine constituents displaying biological activity in vertebrate and invertebrate systems [45–49]. Characterization of these toxins has provided an array of pharmacological tools for the study of glutamate receptors and voltage-dependent calcium channels. Two biologically similar, 48-amino acid peptides, named ω-Agatoxin IVB and IVC, have been isolated from *Agelenopsis* venom [50–52]. These toxins block whole cell Ca^{2+} current in rat Purkinje cells and, along with a related *Agelenopsis* toxin, ω-Aga IVA [49], have been responsible for the discovery and characterization of P-type Ca^{2+} channels. ω-Aga IVB and IVC possess identical amino acid sequences, exist as C-terminal acids and have the same disulfide bond motifs [53]. Reversed-phase HPLC studies confirmed that the two structures are unique entities. Furthermore, dose-response studies revealed an approximately fourfold difference in potency, with ω-Aga IVB blocking a greater portion of P-type Ca^{2+} current more rapidly than ω-Aga IVC [54].

The structural distinction between the two peptides emanated from a series of trypsin and endopeptidase Glu-C digests of ω-Aga IVB and IVC, which provided two different C-terminal fragments [53]. ω-Aga IVC contains an L-serine residue two amino acids from the C-terminus, while ω-Aga IVB reveals a D-serine at this position (Fig. 4). Molecular cloning of the gene encoding both ω-Aga IVB and IVC was accomplished from a cDNA library prepared from *Agelenopsis* venom glands [54]. Translation of this sequence confirmed that the two spider toxins are synthesized from a larger precursor molecule containing both a signal sequence for extracellular transport and a highly acidic propeptide sequence. Furthermore, as observed with the amphibian, mollusc and crustacean-derived peptides, the cDNA encoding the pro-ω-Aga IVB and IVC sequence contains an L-amino acid codon at the D-amino acid site (residue 46).

A major clue to understanding how D-amino acid-containing peptides are synthesized in eukaryotic cells resulted from the discovery of a novel, cofactor-independent serine isomerase found in the venom of the *Agelenopsis aperta* spider [54–57]. Incubation of ω-Aga IVC with crude venom at

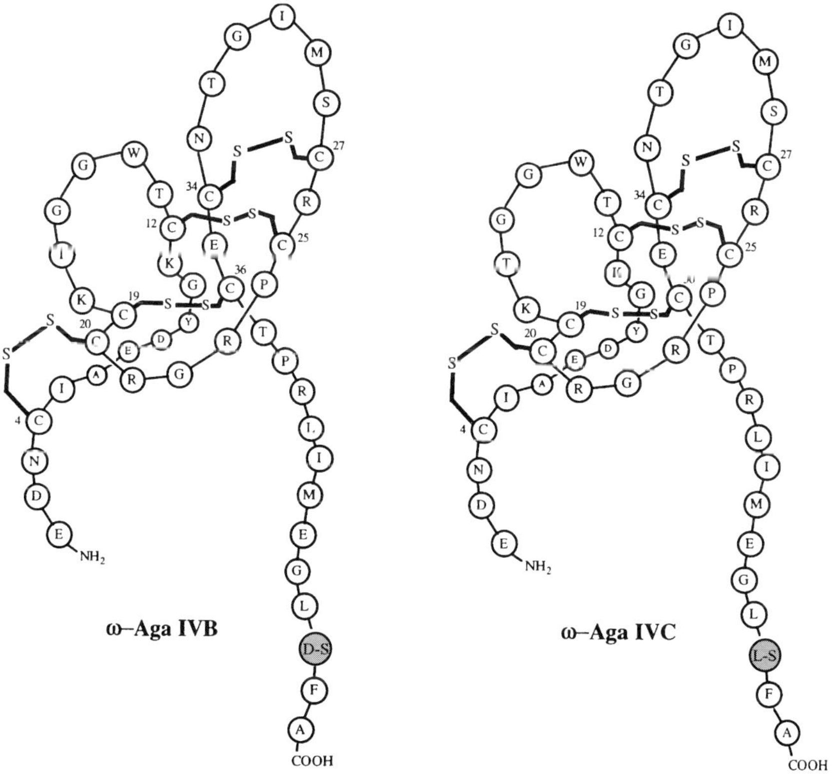

Figure 4. Schematic representation of *Agelenopsis* toxins, ω-Aga IVB and IVC.

37°C resulted in the partial epimerization of the peptide's L-Ser-46 residue. Isomerase activity was reduced by dilution of venom with phosphate-buffered saline (PBS) and was abolished by heating to 100°C. However, isomerization experiments with crude venom were difficult to interpret, since ω-Aga IVC underwent proteolytic cleavage at a rate comparable to that of isomerization. Fractionation of crude venom on a Sephadex G-75 gel column with PBS provided three major high molecular weight components of mass 86.5 kDa, 36.5 kDa and 29.5 kDa, followed by a mixture of lower molecular weight peptides and polyamines [54]. While no enzyme activity was observed with the 36.5-kDa fraction, the 86.5-kDa fraction contained proteolytic activity consistent with a neutral enkephalinase. Proteolysis could be retarded with EDTA or phosphoramidon and reactivated by addition of $ZnCl_2$. The 29.5-kDa fraction contained the desired isomerase activity and was used directly in a series of enzyme experiments. Isomerization of toxin proceeded in both directions, with conversion of ω-Aga IVC to IVB occurring more rapidly than the reverse reaction. In the presence of EDTA, a 50:50 mixture of both isomers was achieved with extended reaction times. Interestingly, isomerization of Ser-28 in the disulfide-rich core of the toxins was not observed, suggesting that the enzyme contains a sequence-specific substrate recognition site.

Characterization of the *A. aperta* isomerase

Enzymes capable of targeting specific amino acids in a peptide chain and altering their absolute stereochemistry have been previously unknown. Nevertheless, their activity has been postulated in recent times to explain the posttranslational conversion of L- to D-amino acids via some form of racemization mechanism. The *Agelenopsis* isomerase is distinguished from other cofactor-independent racemases (e.g., proline [58], mandelate [59, 60], aspartate [61] and glutamate [62]), which operate specifically on free amino acids. Moreover, this same enzymatic activity is not limited to the C-terminal serine residue in ω-agatoxins IVB and IVC.

A minimum substrate recognition site (Leu-Ser-Phe-Ala) for the *Agelenopsis* enzyme was established by examining the enzyme-catalyzed isomerization of small peptide fragments [55] related to the C-terminus of ω-Aga IVC (Tab. 3). The isomerase activity is hardly robust by enzyme standards, with incubation times ranging from hours to days. Peptide fragments were treated with enzyme in the presence of EDTA to minimize proteolytic decomposition. Under these conditions, the enzyme does not target threonine, yet is able to isomerize cysteine, *O*-methyl serine, and alanine amino acid (Xaa) residues in AcNH-Met-Glu-Gly-Leu-Xaa-Phe-Ala-OH; such results rule out the viability of dehydration/rehydration or formaldehyde-mediated hydroxymethylase processes, which would be unique to the hydroxymethyl moiety of serine.

Table 3. Enzyme (ca. 2.3 μM) catalyzed isomerization of small peptide fragments (15.8 μM) related to the C-terminus of ω-Aga IVC in PBS buffer (pH 7.4) for 18 h at 37°C in the presence of 50 μM EDTA

Peptide recognition	% Iso-merized product[a]	Amino acid specificity	% Iso-merized product[a]
Ac-Met Glu Gly Leu Ser Phe Ala-OH	40	Ac-Met Glu Gly Leu Ser Phy Ala-OH	40
Ac-Gly Leu Ser Phe-OH	0	Ac-Met Glu Gly Leu Thr Phe Ala-OH	0
Ac-Glu Gly Leu Ser Phe Ala-OH	29	Ac-Gly Leu L-Ser Phe Ala-OH	50
Ac-Gly Leu Ser Phe Ala-OH	50	Ac-Gly Leu D-Ser Phe Ala-OH	45
Ac-Leu Ser Phe Ala-OH	12	Ac-Gly Leu SerMe Phe Ala-OH	45
Ac-Gly Leu Ser Phe Ala Ala Ala-OH	6 (54)[b]	Ac-Gly Leu Cys Phe Ala-OH	52
Ac-Ala Phe Ser Leu Gly Glu Met-OH	0	Ac-Gly Leu Ala Phe Ala-OH	17 (53)[b]

[a] Percentage of isomerized peptide after 18 h. Isomerized amino acids are underlined.
[b] Percentage of isomerized peptide after 7 days (in parentheses).

The interconversion of AcNH-Gly-Leu-L-Ser-Phe-Ala-OH (**1**) and AcNH-Gly-Leu-D-Ser-Phe-Ala-OH (**2**) (Fig. 5) was examined in greater detail to shed additional light on the *Agelenopsis* isomerase mechanism [55]. k_{cat} for the conversion of **1** to **2** (6.5 sec^{-1}) was found to be ca. 10 times that of the reverse reaction (**2** to **1**). The large difference in k_{cat} and K_m values (ca. 8 mM for the forward reaction and 1 mM for the reverse) between the L → D and D → L directions is intriguing and suggests an asymmetry at the active site, in which one isomer binds more favorably than the other.

The presence (or absence) of a significant deuterium isotope effect in the enzyme-catalyzed isomerization of peptide substrates is an important mechanistic indicator. Epimerization of the serine residue in both isomerizations (L → D and D → L) is significantly retarded by isotopically labeled substrate (^{2}H at the serine α-carbon atom). Although a substantial primary isotope effect on (V/K), where V = V_{max} and K = K_m, is observed in isomerase-catalyzed epimerization for both the L → D and D → L directions [$(V/K)_{^1H}/(V/K)_{^2H}$ of 8.2 in the conversion of **1** to **2** and 4.4 for the reverse process], these values become more illuminating when one compares $V_{^1H}/V_{^2H}$ with values of 31.4 and 4.7 being obtained in the L → D and D → L directions, respectively. This large difference suggests that abstraction of the L-α-proton has a greater impact in determining the rate of the reaction in the L → D direction than abstraction of the D-α-proton in the reverse direction. Mass spectral data indicate that *the solvent-derived*

1 **2**

Figure 5. The *Agelenopsis* isomerase-mediated interconversion of AcNH-Gly-Leu-Ser-Phe-Ala-OH substrates **1** and **2** was examined in detail to probe the enzyme mechanism.

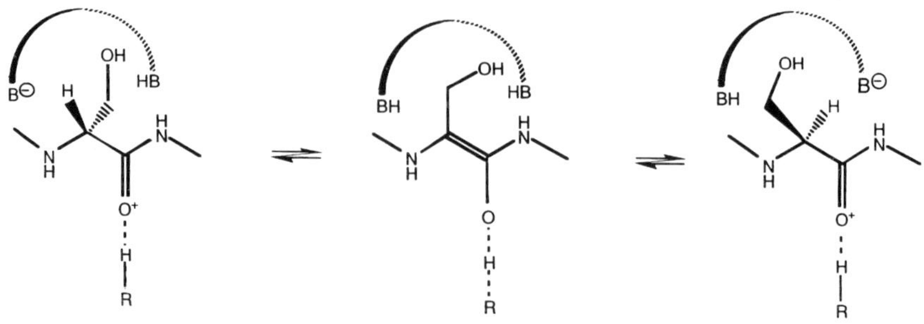

Figure 6. A two-base mechanism is proposed for the peptide isomerase in which abstraction of a proton from one face is concomitant with delivery from the opposite face by the conjugate acid of the second enzymic base. The large difference in K_m and k_{cat} values between the L → D and D → L directions suggests an asymmetry in the active site.

hydrogen atom resides solely with the isomerized product (not substrate) early in the reaction sequence, regardless of whether L- *or* D-*serine-containing substrate is employed.* A substantial primary isotope effect on (V/K) is also observed in isomerase-catalyzed epimerization for both directions when the reaction is carried out in ^2H_2O. Delivery of the α-deuterium atom in the L → D direction is more facile than the reverse D → L direction $[(V/K)_{H_2O}/(V/K)_{^2H_2O}$ of 2.8 vs. 7.6], demonstrating that proton delivery also has a significant impact on the reaction rate. These data point toward a two-base mechanism (Fig. 6), in which proton abstraction from one face is concominant with proton delivery from the opposite face by the conjugate acid of the second enzymic base. Accordingly, the same base responsible for proton abstraction in the L → D conversion and proton delivery (via its conjugate acid) in the D → L conversion has greater difficulty in both transformations than the second base responsible for proton abstraction in the D → L conversion and proton delivery in the L → D conversion.

The asymmetry of the active site, reflected in the isomerization kinetics, raises speculation as to which specific amino acid(s) are responsible for proton abstraction and donation. The *Agelenopsis* isomerase contains a 243-residue heavy chain and an 18-residue light chain connected by a single disulfide bridge [56, 57]. The heavy chain, in turn, contains three intramolecular disulfide bridges and one N-linked oligosaccharide chain with a trimannose core structure. Interestingly, the amino acid sequence of the isomerase is homologous to known serine proteases. Similarity is particularly striking around the regions of the putative catalytic triad of the isomerase (His-46, Asp-96 and Ser-194) and that of other serine proteases (Fig. 7).

How the enzyme is able to remove and redeliver weakly acidic α-protons (pK_a's > 20) in the natural (L) or unnatural (D) configuration is not clear. In recent years, strong low-barrier, high-energy hydrogen bonds [63] (leading to tight binding of enzymes and reactive intermediates) have been proposed for a number of transformations to address the kinetic problems enzymes

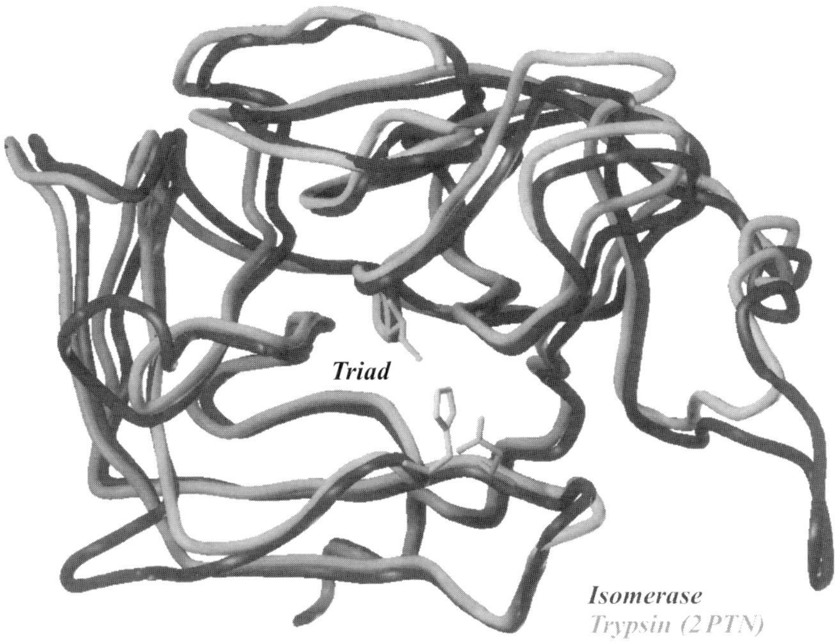

Figure 7. A 3D homology model of the isomerase was built from its published primary sequence and the X-ray structure of trypsin (Brookhaven Protein Data Bank code 2PTN) using the COMPOSER module of Sybyl. (Sybyl Version 6.2.2, Tripos Inc.).

encounter in removing weakly acidic protons (pK_a's > 20) with weakly basic catalysts (pK_a's of conjugate acid < 7) [64, 65]. Formation of a strong hydrogen bond to the amino acid carbonyl moiety would provide a general strategy of the enzyme for reducing ΔG° and $\Delta G^{\ddagger}_{int}$ by promoting carbonyl enolization.

Enzyme activity of the *Agelenopsis* isomerase is severely retarded by the addition of thiol scavenger agents (*N*-ethyl maleimide) [55], consistant with the involvement of active-site cysteines. However, the fact that the amino acid sequence of the enzyme is homologous to other serine proteases and the isomerase is inhibited by serine protease inhibitors (PMSF) further complicates our mechanistic understanding of the enzyme. Progress in elucidating the isomerase's unique interworkings will likely require structural data on enzyme:substrate complexes and efforts directed towards acquiring this evidence are ongoing.

Acknowledgement
We express our gratitude to Dr. James Blake for stimulating discussions and for providing the 3D homology model of the *Agelenopsis* isomerase shown in Figure 7.

References

1. Gevers, W., Kleinkauf, H. and Lipmann, F. (1968) The activation of amino acids for biosynthesis of gramicidin S. *Proc Natl Acad Sci USA* 60:269–276.
2. Kleinkauf, H. and von Döhren, H. (1987) Biosynthesis of peptide antibiotics. *Annu Rev Microbiol* 41:259–289.
3. Jung, G. (1991) Lantibiotics-ribosomal synthesized biologically active polypeptides containing sulfide bridges and α,β-didehydroamino acids. *Angew Chem Int Ed Engl* 30: 1051–1068.
4. Hansen, J.N. (1993) Antibiotics synthesized by posttranslational modification. *Annu Rev Microbiol* 47:535–564.
5. Mor, A., Amiche, M. and Nicolas, P. (1992) Enter a new post-translational modification: D-amino acids in gene-encoded peptides. *Trends Biochem Sci* 17:481–485.
6. Kreil, G. (1994) Peptides containing a D-amino acid from frogs and molluscs. *J Biol Chem* 289:10967–10970.
7. Kreil, G. (1997) D-Amino acids in animal peptides. *Annu Rev Biochem* 66:337–345.
8. Bevins, C.L., Zasloff, M. (1990) Peptides from frog skin. *Annu Rev Biochem* 59:395–414.
9. Lazarus, L.H. and Attila, M. (1993) The toad, ugly and venomous, wears yet a precious jewel in his skin. *Prog Neurobiol* 41:473–507.
10. Erspamer, V., Melchiorri, P., Falconieri-Erspamer, G., Montecucchi, P.C. and de Castiglione R. (1985) *Phyllomedusa* skin: A huge factory and store-house of a variety of active peptides. *Peptides* 6:7–12.
11. Montecucchi, P.C., de Castiglione, R., Piani, S., Gozzini, L. and Erspamer, V. (1981) Amino acid composition and sequence of dermorphin, a novel opiate-like peptide from the skin of *Phyllomedusa sauvagei*. *Int J Pept Protein Res* 17:275–283.
12. Rossi, A.C., de Castiglione, R. and Perseo, G. (1986) Opioid receptor binding profile of selected dermorphin-like peptides. *Peptides* 7:755–759.
13. Broccardo, M., Erspamer, V., Falconieri-Erspamer, G., Improta, G., Linari, G., Melchiorri, P. and Montecucchi, P.C. (1981) Pharmacological data on dermorphins, a new class of potent opioid peptides from amphibian skin. *Br J Pharmac* 73:625–631.
14. Montecucchi, P.C., de Castigline, R., Erspamer, V. (1981) Identification of dermorphin and Hyp^6-dermorphin in skin extracts of Brazilian frog *Phyllomedusa rhodei*. *Int J Pept Protein Res* 17:316–321.
15. Kleinkauf, H. and Döhren, H. (1990) Nonribosomal biosynthesis of peptide antibiotics. *Eur J Biochem* 192:1–15.
16. Kaletta, C., Entian, K.-D. and Jung, G. (1991) Prepeptide sequence of cinnamycin (Ro 09-0198): The first structural gene of a duramycin-type lantibiotic. *Eur J Biochem* 199: 411–415.
17. Richter, K., Egger, R. and Kreil, G. (1987) D-Alanine in the frog skin peptide dermorphin is derived from L-alanine in the precursor. *Science* 238:200–202.
18. Bradbury, A. and Smyth, D.G. (1991) Peptide amidation. *Trends Biochem Sci* 16:112–115
19. Kreil, G., Barra, D., Simaco, M., Erspamer, V. and Falconieri, P. (1989) Deltorphin, a novel amphibian skin peptide with high selectivity and affinity for δ opioid receptors. *Eur J Pharmacol* 162:123–128.
20. Lazarus, L.H., Wilson, W.E., de Castiglione, R. and Guglietta, G. (1989) Dermorphin gene sequence peptide with high affinity and selectivity for δ-opioid receptors. *J Biol Chem* 264:3047–3050.
21. Mor, A., Delfour, A., Sagan, S., Amiche, M., Pradelles, P., Rossier, J. and Nicolas, P. (1989) Isolation of dermenkephalin from amphibian skin, a high-affinity δ-selective opioid heptapeptide containing a D-amino acid residue. *FEBS Lett* 255:269–274.
22. Sagan, S., Amiche, M., Delfour, A., Mor, A., Camus, A. and Nicolas, P. (1989) Molecular determinants of receptor affinity and selectivity of the natural Δ-opioid agonist, dermenkephalin. *J Biol Chem* 264:17100–17106.
23. Erspamer, V., Melchiorri, P., Falconieri-Erspamer, G., Negri, L., Corsi, R., Severini, C., Barra, D., Simmaco, M. and Kreil, G. (1989) Deltorphins: A family of naturally occurring peptides with high affinity and selectivity for δ opioid binding sites. *Proc Natl Acad Sci USA* 86:5188–5192.

24. Erspamer, V. (1992) The opioid peptides of the amphibian skin. *Int J Dev Neurosci* 10: 3–30.
25. Barra, D., Mignogna, G., Simmaco, M., Pucci, P., Severini, C., Falconieri-Erspamer, G., Negri, L. and Erspamer, V. (1994) [D-Leu2]deltorphin, a 17 amino acid opioid peptide from the skin of the Brazilian hylid frog, *Phyllomedusa burmeisteri*. *Peptides* 15:199–202.
26. Mor, A., Delfour, A. and Nicolas, P. (1991) Identification of a D-alanine-containing polypeptide precursor for the peptide opioid, dermorphin. *J Biol Chem* 266:6264–6270.
27. Simmaco, M., Barra, D., Chiarini, F., Noviello, L., Melchiorri, P., Kreil, G. and Richter, K. (1991) A family of bombinin-related peptides from the skin of *Bombina variegata*. *Eur J Biochem* 199:217–222.
28. Mignogna, G., Simmaco, M., Kreil, G. and Barra, D. (1993) Antibacterial and haemolytic peptides containing D-alloisoleucine from the skin of *Bombina variegata*. *EMBO J* 12:4829–4832.
29. Kamatani, Y., Minakata, H., Kenny, P.T.M., Iwashita, T., Watanabe, K., Funase, K., Ping Sun, X., Yongsiri, A., Kim, K.H. and Novales-Li, P. et al. (1989) Achatin-I, an endogenous neuroexcitatory tetrapeptide from *Achatina fulica* Ferussac containing a D-amino acid residue. *Biochem Biophys Res Commun* 160:1015–1020.
30. Fujimoto, K., Kubota, I., Yasuda-Kamatani, Y., Minakata, H. Nomoto, K., Yoshida, M., Harada, A., Muneoka, Y. and Kobayashi, M. (1991) Purification of achatin-I from the atria of the African giant snail, *Achatina fulica*, and its possible function. *Biochem Biophys Res Commun* 177:847–853.
31. Kamatani, Y., Minakata, H., Iwashita, T., Nomoto, K., In, K., Doi, M., Ishida, T. (1990) Molecular conformation of achatin-I, an endogenous neuropeptide containing D-amino acid residue. *FEBS Lett* 276:95–97.
32. Ishida, T., In, Y., Doi, M., Inoue, M., Yasuda-Kamatani, Y., Minakata, H., Iwashita, T., Nomoto, K. (1992) Crystal structure and molecular conformation of achatin-1 (H-Gly-D-Phe-Ala-Asp-OH), an endogenous neuropeptide containing a D-amino acid residue. *Int J Peptide Protein Res* 39:258–264.
33. Ishida, T., In, Y., Inoue, M., Yasuda-Kamatani, Y., Minakata, H., Iwashita, T. and Nomoto, K. (1992) Effect of the D-Phe^2 residue on molecular conformation of an endogenous neuropeptide achatin-I. *FEBS Lett* 307:253–256.
34. Ohta, N., Kubota, I., Takao, T., Shimonishi, Y., Yasuda-Kamatani, Y., Minakata, H., Nomoto, K., Muneoka, Y. and Kobayashi, M. (1991) Fulicin, a novel neuropeptide containing a D-amino acid residue isolated from the ganglia of *Achatina fulica*. *Biochem Biophys Res Commun* 178:486–493.
35. Fujita, K., Minakata, H., Nomoto, K., Furukawa, Y. and Kobayashi, M. (1995) Structure-activity relations of fulicin, a peptide containing a D-amino acid residue. *Peptides* 16:565–568.
36. Yusuda-Kamatani, Y., Nakamura, M., Minakata, H., Nomoto, K. and Sakiyama, F. (1995) A novel cDNA sequence encoding the precursor of the D-amino acid-containing neuropeptide fulicin and multiple α-amidated neuropeptides from *Achatina fulica*. *J Neurochem* 64:2248–2255.
37. Yasuda-Kamatani, Y., Kobayashi, M., Yasuda, A., Fujita, T., Minakata, H., Nomoto, K., Nakamura, M. and Sakiyama, F. (1997) A novel D-amino; acid-containing peptide, fulyal, coexits with fulicin gene-related peptides in *Achatina atria*. *Peptides* 18:347–354.
38. Fujisawa, Y., Ikeda, T., Nomoto, K., Yasuda-Kamatani, Y., Minakata, H., Kenny, P.T.M., Kubota I. and Muneoka Y. (1992) The FMRFamide-related decapeptide of *Mytilus* contains a D-amino acid residue. *Comp Biochem Physiol* 102C:91–95.
39. Yasuda, A., Yasuda, Y., Fujita, T. and Naya, Y. (1994) Characterization of crustacean hyperglycemic hormone from the crayfish (*Procambarus clarkii*): multiplicity of molecular forms by stereoinversion and diverse functions. *Gen Comp Endocrinol* 95:387–398.
40. De Kleijn, D.P.V., Janssen, K.P.C., Martens, G.J.M. and Van Herp, F. (1994) Cloning and expression of two crustacean hyperglycemic hormone mRNAs in the eyestalk of the crayfish *Orconectes limosus*. *Eur J Biochem* 224:623–629.
41. Aguilar, M.B., Soyez, D., Falchetto, R., Arnott, D., Shabanowitz, J., Hunt, D.F., Huberman A. (1995) Amino acid sequence of the minor isomorph of the crustacean hyperglycemic hormone (CHH-II) of the Mexican crayfish *Procambarus bouvieri* (Ortmann): presence of a D-amino acid. *Peptides* 16:1375–1383.

42. Tensen, C.P., De Kleijn, D.P.V. and Van Herp, F. (1991) Cloning and sequence analysis of cDNA encoding two crustacean hyperglycemic hormones from the lobster *Homarus americanus*. *Eur J Biochem* 200: 103–106.
43. Soyez, D., Van Herp, F., Rossier, J., Le Caer, J.-P., Tensen, C.P. and Lafont, R. (1994) Evidence for a conformational polymorphism of invertebrate neurohormones. D-Amino acid residue in crustacean hyperglycemic peptides. *J Biol Chem* 269: 18295–18298.
44. Jimenéz, E.C., Olivera, B.M., Gray, W.R. and Cruz, L.J. (1996) Contryphan is a D-tryptophan-containing *Conus* peptide. *J Biol Chem* 271: 28002–28005.
45. Jasys, V.J., Kelbaugh, P.R., Nason, D.M., Phillips, D., Rosnack, K.J., Forman, J.T., Saccomano N.A., Stroh, J.G. and Volkmann, R.A. (1992) Novel quaternary ammonium salt-containing polyamines from the *Agelenopsis aperta* funnel-web spider. *J Org Chem* 57: 1814–1820.
46. Parks, T.N., Mueller, A.L., Artman, L.D., Albensi, B.C., Nemeth, E.F., Jackson, H., Jasys, V.J., Saccomano, N.A. and Volkmann, R.A. (1991) Aryl amine toxins from funnel-web spider (*Agelenopsis aperta*) venom antagonize *N*-methyl-D-aspartate receptor function in mammalian brain. *J Biol Chem* 266: 21523–21529.
47. Saccomano, N.A. and Volkmann, R.A. (1993) Polyamines and polypeptides useful as antagonists of excitatory amino acid neurotransmitters and/or as blockers of calcium channels. US Patent 5,227,397, filed 17 July, 1990.
48. Skinner, W.S., Adams, M.E., Quistad, G.B., Kataoka, H., Cesarin, B.J., Enderlin, F.E. and Schooley, D.A. (1989) Purification and characterization of two classes of neurotoxins from the funnel web spider, *Agelenopsis aperta*. *J Biol Chem* 264: 2150–2155.
49. Mintz, I.M., Venema, V.J, Swiderek, K.M., Lee, T.D., Bean, B.P. and Adams, M.E. (1992) P-type calcium channels blocked by spider toxin ω-Aga IVA. *Nature* 355: 827–829.
50. Yu, H, Rosen, M.K., Saccomano, N.A., Phillips, D., Volkmann, R.A. and Schreiber, S.L. (1993) Sequential assignment and structure determination of spider toxin ω-Aga-IVB. *Biochemistry* 32: 13123–13129.
51. Adams, M.E., Mintz, I.M., Reily, M.D., Thanabal, V. and Bean, B.P. (1993) Structure and properties of ω-agatoxin IVB, a new antagonist of the P-type calcium channels. *Mol Pharmacol* 44: 681–688.
52. Teramoto, T., Kuwada, M., Niidome, T., Sawada, K., Nishizawa, Y. and Katayama, K. (1993) A novel peptide from the funnel web spider, ω-Aga-TK, selectively blocks P-type calcium channels. *Biochem Biophys Res Commun* 196: 134–140.
53. Heck, S.D., Kelbaugh, P.R., Kelly, M.E., Thadeio, P.F., Saccomano, N.A., Stroh, J.G. and Volkmann, R.A. (1994) Disulfide bond assignment of ω-agatoxins IVB and IVC: discovery of a D-serine residue in ω-agatoxin IVB. *J Amer Chem Soc* 116: 10426–10436.
54. Heck, S.D., Siok, C.J., Krapcho, K.J., Kelbaugh, P.R., Thadeio, P.F., Welch, M.J., Williams, R.D., Ganong, A.H., Kelly, M.E. and Lanzetti, A.J. et al. (1994) Functional consequences of post-translational isomerization of Ser46 in a calcium channel toxin. *Science* 266: 1065–1068.
55. Heck, S.D., Faraci, W.S., Kelbaugh, P.R., Saccomano, N.A., Thadeio, P.F. and Volkmann, R.A. (1996) Posttranslational amino acid epimerization: enzyme-catalyzed isomerization of amino acid residues in peptide chains. *Proc Natl Acad Sci USA* 93: 4036–4039.
56. Shikata, Y., Watanabe, T., Teramoto, T., Inoue, A., Kawakami, Y., Nishizawa, Y., Katayama, K. and Kuwada, M. (1996) Structure and function of a novel peptide isomerase from funnel web spider venom. *Pept Chem*. Volume date 1995, 165–168.
57. Shikata, Y., Watanabe, T., Teramoto, T., Inoue, A., Kawakami, Y., Nishizawa, Y., Katayama, K. and Kuwada, M. (1995) Isolation and characterization of a peptide isomerase from funnel web spider venom. *J Biol Chem* 270: 16719–16723.
58. Rudnick, G. and Abeles, R.H. (1975) Reaction mechanism and structure of the active site of proline racemase. *Biochemistry* 14: 4515–4522.
59. Neidhart, D.J., Howell, P.L., Petsko, G.A., Powers, V.M., Li, R., Kenyon, G.L. and Gerlt, J.A. (1991) Mechanism of the reaction catalyzed by mandelate racemase. 2. Crystal structure of mandelate racemase at 2.5 Å resolution: identification of the active site and possible catalytic residues. *Biochemistry* 30: 9264–9273.
60. Babbit, P.C., Mrachko, G.T., Hasson, M.S., Huisman, G.W., Kolter, R., Ringe, D., Petsko, G.A., Kenyon, G.L. and Gerlt, J.A. (1995) A functionally diverse enzyme superfamily that abstracts the α-protons of carboxylic acids. *Science* 267: 1159–1161.

61. Yamauchi, T., Choi, S.-Y., Okada, H., Yohda, M., Kumagai, H., Esaki, N. and Soda, K. (1992) Properties of aspartate racemase, a pyridoxal 5′-phosphate-independent amino acid racemase. *J Biol Chem* 267:18361–18364.
62. Gallo, K.A. and Knowles, J.R. (1992) Purification, cloning, and co-factor independence of glutamate racemase from *Lactobacillus*. *Biochemistry* 32:3981–3990.
63. Cleland, W.W. and Kreevoy, M.M. (1994) Low-barrier hydrogen bonds and enzymic catalysis. *Science* 264:1887–1890.
64. Gerlt, J.A. and Gassman, P.G. (1993) Understanding of the rates of certain enzyme-catalyzed reactions: proton abstraction from carbon acids, acyl-transfer reactions and displacement reactions of phosphodiesters. *Biochemistry* 32:11943–11952.
65. Gerlt, J.A., Kozarich, J.W., Kenyon, G.L. and Gassman, P.G. (1991) Electrophilic catalysis can explain the unexpected acidity of carbon acids in enzyme-catalyzed reactions. *J Am Chem Soc* 113:9667–9669.

Appearance of D-amino acids during aging

D-Amino Acids in Sequences of Secreted Peptides of Multicellular Organisms
ed. by P. Jollès

Appearance of D-amino acids during aging: D-Amino acids in tumor proteins

G.H. Fisher

Department of Chemistry, Barry University, 11300 Northeast Second Avenue, Miami, FL 33161, USA

Summary. In 1939 Kögl and Erxleben [1–4] reported that tumor proteins contain appreciable amounts of D-amino acids, specifically glutamic acid, valine, leucine and lysine, implying that both the initiation and autonomous character of tumors depends on the formation and maintenance of these D-amino acids in the cell proteins. This postulate remained highly controversial for over 10 years, during which time several papers both supporting and refuting this hypothesis were published. The dispute existed almost entirely between Kögl, a vigorous and able protagonist at the University of Utrecht, Netherlands, and an impressive array of equally vigorous and able dissenters in the United Kingdom and Germany. An excellent review of both sides of this controversy was written by Miller in 1950 [5]. After many years and much effort the controversy then seemed to be put to rest. However, more than 40 years later the development of much more refined analytical techniques for the resolution and detection of amino acid enantiomers provided more definitive evidence that D-amino acids are not common to all tumor tissues and probably are not integral to the cancer process. This is not surprising when one considers that a tumor consists of fast-growing cells. Thus, there would not be sufficient time for any L-amino acid to racemize to the D isomer. Some D-amino acids may originate in foods consumed, but it is uncertain whether enzyme systems are able to incorporate D-amino acids into tumor proteins during growth. Nevertheless, if significant levels of D-amino acids were to be found in tumor proteins, the implications could be far-reaching. Confirmation of the presence of D-amino acids at any concentration in tumors would provide new insights into the mechanism for autogenesis and maintenance of tumors.

Early work: 1940s and 1950s

Kögl's hypothesis and discoveries

The basic hypothesis of Kögl was that tumor proteins are characterized by the presence of certain amino acids in the D form, particularly D-glutamic acid (D-Glu). He further inferred that the initiation and autonomous character of tumor cells are dependent on the formation and maintenance of these D-amino acids in cell proteins. This seemed to be a very attractive hypothesis because it provided a basic premise in relatively definite chemical terms of how a cell might very subtly become and remain independent of the controls that govern normal growth.

In the initial work of Kögl and Erxleben [1–4], crude proteins extracted from normal and malignant tissues were refluxed in concentrated HCl for 7 h, and samples of about 15 amino acid hydrochlorides were obtained from the hydrolysates in relatively pure state. The amino acid hydrochlorides from the normal or tumor hydrolysates were crystallized in the cold after saturation with HCl gas. Such preparations generally had the correct C,H,N elementary composition after one recrystallization. The percent of D enantiomer was calculated from the specific optical rotation of the samples. Analysis of several normal tissues indicated that only L-amino acids were present, whereas four amino acids from the tumor proteins appeared to be racemized to a greater extent than is formed by partial racemization under similar conditions in pure solution. Leucine, lysine and valine contained about 3–4% of the D form, while glutamic acid samples from several tumors were calculated to contain as much as 32–45% of the D form. These and later extensive analyses by Kögl et al. [6–8] also indicated that the amount of D-Glu present was roughly proportional to the malignancy of the tumor in question.

The controversy

Because of the latter finding and the apparent ease with which glutamic acid hydrochloride could be isolated in pure state from protein hydrolysates, the controversy that soon developed centered on glutamic acid alone. As expected, many attempts were made to repeat these important findings, and within a short time several notes [9–12] appeared which confirmed Kögl's results. Several other papers were also published from many laboratories which contradicted Kögl, using either his methods or other isolation procedures [13–26]. Kögl and his group published at least 10 more original papers in the 1940s dealing with further analyses and different attacks on the dissenters, and in each of these their initial results were repeatedly confirmed.

Much of the difference between these groups centered on factors such as the solubility of the racemic D,L-glutamate salts compared with that of either pure enantiomer and whether the first crop of the glutamate hydrochloride was more or less racemized than succeeding crops. Although opponents could also obtain first crops with high degrees of racemization, the disagreement at this point centered on Kögl's assumption that such preparations were representative of the total glutamic acid present. Dissenters reasoned that it is not permissible to calculate the amount of D-Glu present solely from the specific rotation of the first fraction; instead, all the fractions must be considered. When this was done, the amount of D-Glu present in tumor protein hydrolysates was less than 3% of the total glutamic acid present [9, 13–18, 20, 27]. Kögl et al. [28] attempted to answer this criticism by showing that the second crop of Glu · HCl had only a slightly lower specific rotation than the first crop, while the data of Chibnall et al. [14–16] and of Woodward et al. [27] showed that the specific rotations of successive crops either approach or do not deviate essentially from the rotation of the L enantiomer.

While Kögl et al. could not detect any D-Glu in hydrolysates of normal tissue, other workers found the same low percentage of D-Glu in hydrolysates of both normal and malignant tissues. A similar finding was made for aspartic acid [16], which Kögl et al. [1–4, 8] claimed was present only in the L form in both normal and tumor tissues. This is surprising in light of the fact that the rate of racemization of aspartic acid is greater than that of glutamic acid [29]. The general feeling among the opposition was that the small amounts of D-Glu found in both normal and malignant tissues resulted from the heat-induced racemization of glutamic acid during the hydrolysis process, since the observed racemization of L-Glu in pure solution was of the same order of magnitude [15, 21, 26, 30, 31].

In retrospect, it appears that too little attention was paid by the Utrecht school and the dissenters to the criteria of purity necessary to demonstrate that the optical rotation of any given sample of glutamic acid is actually a measure of the D-Glu present. NaCl and NH_4Cl and other amino acids are difficult to remove by recrystallization from glutamic acid hydrochloride, particularly when the recrystallization is carried out with the small weight losses (10% or less) recorded and recommended in the Kögl papers. Such contaminants would obviously alter the specific optical rotation, and NaCl and NH_4Cl would escape detection in the usual ash determination. While C,H,N and ash determinations were most frequently used as criteria of purity, it is apparent from the work of Klingmüller [22] and Abderhalden [13] that further criteria must be applied to eliminate the possibility that contaminating amino acids contribute to the observed optical rotations. Furthermore, when Kögl et al. obtained large yields of the glutamic hydrochloride, they did not prove that the low rotations of these preparations could not have been due to contaminating amino acids and salts. In any case, the isolation in one instance of a relatively large amount of D-glutamate (10.3% of protein weight as Glu · HCl, 44.5% D form) from the very malignant Brown-Pearce rabbit tumor [3] remains to confront Kögl's opposition, although it is clear that this one example does not prove Kögl's general thesis.

Other methods of detecting D-amino acids in tumors

Lipmann et al. [32] employed D-amino acid oxidase to test tumor protein hydrolysates for D-amino acids. While they recognized that D-Glu is attacked slowly by this enzyme, they were able to recover D-Glu added as the racemate to hydrolysates of normal tissue, but in no case could they detect more than 1–3% of D-acid-nitrogen in the total nitrogen of the hydrolysates of tumors. However, when Kögl, Herken and Erxleben [33] employed D-amino acid oxidase, they could not recover added D-Glu from protein hydrolysates and concluded that it could not be used in such complex mixtures, particularly for an amino acid as slowly oxidized as glutamic acid.

In 1968 Berezov [34] studied deamination of L and D isomers of dicarboxylic acids in malignant tumors. Deamination of D isomers of glutamic and aspartic acids was absent in malignant tumors and in control tissues of man and animals. L isomers were deaminated quickly in normal tissues only. In tumors, the speed of deamination was 1/30 to 1/5 times that of controls. On the basis of these findings it was concluded that there are no specific oxidases of D-amino acids in tumors or that they are not in active state as far as deamination of D isomers of Glu and Asp are concerned, inferring that D-Glu and D-Asp may be present in tumor proteins.

In 1949 Boulanger and Osteux [35] isolated glutamic acid from hydrolysates of proteins by adsorption on acidified alumina and elution with dilute acetic acid. Following destruction of the L-Glu by bacterial L-glutamic acid decarboxylase, the preparations were chromatographed on paper to obtain the concentrations of D-Glu remaining. They found 0.6 to 2.4% of D-Glu in the total glutamic acid from a variety of *both* normal and tumor proteins.

Feeding studies with animals

Despite the many contradictions that resulted between the Utrecht group and others, Kögl and colleagues in the late 1940s published additional studies with animals to prove their supposition that tumor proteins contain D-Glu. Use was made of the observation that when D-Glu was fed to rats, about 75% of single doses of D-Glu appeared in the urine as D-α-pyrrolidone carboxylic acid [36], while L-Glu did not yield the corresponding lactam in the urine. Kögl et al. [37] reported in 1948 that after feeding cooked rat tumors to a dog, a small amount of pure D-α-pyrrolidone carboxylic acid could be isolated from the urine. However, the amount obtained was estimated to be about 10% of theoretical based on their usual optical analyses. The same dog was later fed beef protein, and the urine was extracted as before, but no attempt was made to isolate any α-pyrrolidone carboxylic acid. Instead, a sample of the D-lactam labeled with deuterium was added to the extract and then reisolated. This material did not show a lowered isotope content, so it was concluded that no D-α-pyrrolidone carboxylic acid was present in the urine obtained after feeding beef protein to the dog. The D-lactam was also obtained from the urine of rats fed with similar tumors.

These feeding experiments were repeated by Hillmann et al. in 1954 [38]. After feeding 1500 g of human liver metastases to rats, 190 mg of pure D-α-pyrrolidone carboxylic acid was obtained from their urine. However, other feeding experiments with two human liver metastases did not yield any crystallized D-pyrrolidonic acid. After feeding rats with human liver, beef, bread or grain equal in amount to the tumor tissue used in these experiments, they were never able to isolate pyrrolidonic acid; but after direct hydrolysis of the ethyl acetate extract of the urine, the presence of a small amount of D-Glu was shown by paper chromatography. In order to determine if this D-Glu arose from ordinary food proteins, the urine of starved rats was analyzed; D-Glu was also shown to be present. The same result was obtained with human urine. Since D-Glu was found in the urine of starving rats and humans, it may well be that the presence of unphysiological D-amino acids has to be taken into account in ordinary metabolism.

In 1951, Mayer [39] presented a different interpretation of the facts known up to that point. Foods containing albuminoids also have small amounts of D-amino acids. Normally the harmful effects of these D-amino acids are neutralized either by natural elimination or by deamination in persons with normal hepatic and renal functions. However, in persons having hepatic and renal malfunction or in those with a precancerous focus, D-Glu accumulates in certain points and changes the stereochemical structure of certain proteins which act in tissue formation. In fact, he said, the appearance of neoplasms may be due to "the action of amino acids of abnormal stereochemical type."

Recent studies (1980s – present)

The interest of D-amino acids in tumors seemed to die down after the 1950s with the exception of the isolated report of Berezov in 1968 [34] mentioned previously. However, technological refinements developed recently have provided analytical methods for the detection of amino acid enantiomers with far more accuracy and sensitivity than those used by the earlier investigators.

Chiral GC analysis

In 1980, Smith and Wonnacott [40] described an improved method for the resolution of enantiomeric isopropyl esters of *N*-trifluoroacetyl-α-amino acids using mixed chiral phases on a stainless-steel capillary gas chromatographic column. Using this analysis technique, Stroud and Smith [41] in 1984 analyzed four different tumor types from three species. In all the samples, Ala, Val, Ile, Leu, Pro, Met, Phe, Asp and Glu were identified and resolved. After correction for the process-induced racemization, no D isomers were found for any of the amino acids. Special concern was given to Val, Leu and Glu, which had previously been reported to occur in both D and L forms in tumor proteins. Careful examination, with gas chromatography/mass spectrometry (GC/MS), of the area of the D-isomer elution of these amino acids showed that absolutely none was present.

Current author's work

The current author, Fisher et al. [42] have also analyzed several sets of tumors and normal control tissues for the presence of D-aspartate and D-glutamate. Various human tumors and control tissues were homogenized with in cold 10% trichloroacetic acid (TCA), centrifuged and washed with doubly distilled water to remove the acid-soluble free amino acids into solution while precipitating most of the proteins. The protein pellet was hydrolyzed for 6 h at 100°C in 6 N doubly distilled HCl in glass tubes sealed under vacuum. The hydrolysate was evaporated by rotary evaporation and partially purified by passage through a C-18 Sep-Pak clean-up cartridge. Asp and Glu were isolated by anion exchange chromatography on an AG 1-X8 column, washed first with water to remove the basic and neutral amino acids and then with 0.5 M acetic acid to elute Asp and Glu.

The D/L-amino acid ratios were determined by capillary gas chromatography (GC) using the chiral column method of Smith and Wonnacott [40], and some analyses were double-checked by high performance liquid chromatography (HPLC), using a technique developed by Aswad [43] for separating the diastereomeric amino acid derivatives produced by reaction of the D/L-amino acid mixture with *o*-phthaldialdehyde (OPA) in the presence of *N*-acetyl-L-cysteine to form a pair of fluorescent diastereomeric *N*-alkyl-2-thioalkyl isoindoles. Quantitative determination of the percent D-Asp and D-Glu was accomplished by area measurement of the respective GC or HPLC peaks using a computing integrator. Standards of known D/L-ratios were run as controls. D/L-ratios were corrected for hydrolysis-induced racemization of L to D enantiomers.

The percent D-Asp and D-Glu found in these samples are given in Table 1. In all samples, the average percent of D isomer found is ≤ 1%. The levels of D-Asp are slightly higher than those of D-Glu, since Asp racemizes faster than Glu (tumors: 0.72% D-Asp, 0.61% D-Glu; controls: 0.94% D-Asp, 0.35% D-Glu). Most tumors contain less D-Asp than controls, whereas nearly half of the tumors contain more D-Glu than controls. However, within the limits of experimental error, there is no statistically significant difference in the levels of these D-amino acids between the tumors and normal tissues.

D-Amino acid oxidase in tumors

D-Amino acid oxidase (DAO, EC 1.4.3.3) is a flavoprotein enzyme which catalyzes oxidative deamination of D-amino acids to α-keto acids. The enzyme is found widespread is many animal tissues such as kidney, liver, brain, leucocytes and so on. The physiological role of the enzyme is to remove D-amino acids from dysfunctional systems. Thus, the absence or decrease of DAO could allow the accumulation of D-amino acids. Experi-

Table 1. Percent D-Asp and D-Glu in tumors and control tissues

Tissue	Type of tumor	% D-Asp*		% D-Glu*	
		Tumor	Control	Tumor	Control
Lung	Adenoid cystic carcinoma	0.76 ± 0.08	1.20 ± 0.02	0.28 ± 0.006	0.27 ± 0.02
	Adenosquamous carcinoma	1.00 ± 0.07	0.90 ± 0.11	1.05 ± 0.28	0.50 ± 0.08
Neck	Squamous cell carcinoma	0.54 ± 0.08	0.04 ± 0.06	0.67 ± 0.07	0.24 ± 0.07
Kidney	Adenocarcinoma	0.59 ± 0.36	0.63 ± 0.08	0.40 ± 0.08	0.42 ± 0.05
	Transitional cell carcinoma	0.52 ± 0.04		0.42 ± 0.02	
Colon	Adenocarcinoma	0.46 ± 0.08	1.05 ± 0.08	0.37 ± 0.07	0.47 ± 0.04
Ovary	Adenocarcinoma	1.05 ± 0.14	1.72 ± 0.07	0.77 ± 0.08	0.47 ± 0.15
	Possible adeno-carcinoma of adnexa	0.64 ± 0.08		0.30 ± 0.20	
Uterus	Adenocarcinoma	1.02 ± 0.07	0.83 ± 0.16	< 0.01	< 0.01
	Malignant mixed mullerian adeno-carcinoma	0.58 ± 0.10	1.37 ± 0.09	0.62 ± 0.18	0.34 ± 0.04
Breast	Infiltrating ductal carcinoma	0.90 ± 0.12	1.14 ± 0.17	0.03 ± 0.05	0.35 ± 0.61
Tongue	Squamous cell carcinoma	0.65 ± 0.12	0.75 ± 0.10	1.05 ± 0.20	0.42 ± 0.09
Parotid	Pleomorphic adeno-carcinoma	0.66 ± 0.11	0.73 ± 0.14	1.93 ± 0.30	0.36 ± 0.07
	Average	0.72 ± 0.20	0.94 ± 0.44	0.61 ± 0.52	0.35 ± 0.14

$*\%\text{D} = \frac{\text{D}}{\text{D} + \text{L}} \times 100$. D/L-Asp ratios are corrected for hydrolysis-induced racemization.

ments have been carried out to determine if the difference in rates of cell proliferation between normal and neoplastic cells may be related to altered levels of this oxidative enzyme. There have been only a couple of reports correlating the presence (or absence) of DAO with tumors. In 1980, Sun and Cederbaum [44] found that the levels of DAO were either undetected or 12-fold lower in hepatocellular carcinoma and precancerous tissue than in normal liver homogenates. Similarly, in 1990, Abdellatif et al. [45] reported that rats fed with cancer-causing agents (nafenopin or per-fluorooctanoic acid) had higher incidences of hepatocellular carcinoma and significantly depressed levels of D-amino acid oxidase and glycolate oxidase than control animals. These findings could imply that D-amino acids might be able to accumulate in tumors, since enzymes for their removal are absent or at lower levels. However, this still does not prove the existence of D-amino acids in tumors.

Uptake of D-amino acids in tumors

Recently several experiments have been undertaken to examine the infusion of radiolabeled D- and L-amino acids into tumors for use as radioactive nuclear imaging agents. Goto et al. [46–50] reported that various D-amino acids were taken up several times more effectively than the corresponding L isomers into several tumors tested *in vivo*. The radioactivity of D-amino acids tested (Ala, Leu, Met, Trp) was found most highly in tumor cells and pancreas, and the radioactivity in tumor cells was several times higher than that of the L forms. In contrast, Schober et al. [51] reported nonselective transport of [^{11}C-methyl]-L- or D-methionine into malignant glioma brain cells. Nevertheless, these findings suggest the potential utility of gamma-emitter radiolabeled D-amino acids for the detection of some cancers by X-ray computed tomography (CT) or position emission tomography (PET) imaging.

Conclusions

Even though it is generally believed that the rates of amino acid racemization under various hydrolysis conditions are sufficient to account for the low levels of D-amino acids found in proteins hydrolyzates [31, 32], these findings do not exclude other possibilities. Some as yet unknown tumor proteins may actually contain low levels of D-amino acids, or the L-amino acid residues in certain tumor proteins may be in sequences which are more easily racemized upon hydrolysis. Nor is it inconceivable that amino acid racemases might play a role. On the other hand, one could speculate on the mode of action of carcinogens that might induce racemization within peptide chains or that they might cause the formation of easily racemizable sequences of amino acids in tumor proteins. Some D-amino acids may originate in foods consumed [52, 53], but it is uncertain whether enzyme systems are able to incorporate D-amino acids into tumor proteins during growth.

This author's own results dispute the original hypothesis of Kögl and Erxleben that tumors contain significant amounts of D-amino acids, and we confirm the findings of other recent workers who also did not find any D-amino acids in tumors. This is not surprising when one considers that a tumor consists of fast-growing cells. Thus, there would not be sufficient time for any L-amino acids to racemize to the D-isomer. Nevertheless, if significant levels of D-amino acids were to be found in tumor proteins, the implications could be far-reaching. Confirmation of the presence of altered amino acids at any concentration in tumors would provide new insights into the mechanism for autogenesis and maintenance of tumors.

References

1. Kögl, F. (1939) The glutamic acid of tumour proteins. *Nature* 144:111.
2. Kögl, F. and Erxleben, H. (1939) Zur Ätiologie der malignen Tumoren. *Z Physiol Chem* 258:57–95.
3. Kögl, F., Erxleben, H. and Akkerman, A.M. (1939) Isolierung der Glutaminsäure aus Tumorproteinen. *Z Physiol Chem* 261:141–153.
4. Kögl, F. and Erxleben, H. (1939) Zur Stereochemie der Proteine von Myomen und anderen Geschwülsten. *Z Physiol Chem* 261:154–157.
5. Miller, J.A. (1950) Do tumor proteins contain D-amino acids? *Cancer Res* 10:65–72.
6. Erxleben, H. and Herken, H. (1940) Beiträge zur stereochemischen Analyse von Proteinen des Lymphogranuloma und anderen pathologischer Gewebe. *Z Physiol Chem* 264:240–250.
7. Kögl, F. (1940) Zur Ätiologie der Tumoren. *Z Krebsforsch* 49:291–311.
8. Kögl, F. (1942) Bemerkung zu neueren Arbeiten über die Chemie der Tumoren. *Naturwis* 30:46.
9. Arnow, L.E. and Opsahl, J.C. (1939) The configuration of the glutamic acid of adenocarcinoma protein. *Science* 90:257–258.
10. Arnow, L.E., Opsahl, J.C. and Watson, C.J. (1940) Configuration of the glutamic acid isolated from subacute lymphatic leukemic tissue proteins. *Soc Exper Biol Med* 43:765–766.
11. Ottawa, H. (1940) Zur Frage der optische Aktivität der aus Tumoreiweiss isolierten Glutaminsäuer. *Z Krebsforsch* 49:677–678.
12. White, J. and White, F.R. (1939) On the nature of the glutamic acid of malignant tumor tissues. *J Biol Chem* 130:435–436.
13. Abderhalden, E. (1942) Prüfung auf das Vorkommen von *d*-Glutaminsäure an Eiweiss von malignen Tumoren und von normalen Organen. *Z Physiol Chem* 275:135–154.
14. Chibnall, A.C., Rees, M.W., Tristram, G.R., Williams, E.F. and Boyland, E. (1939) The glutamic acid of tumour proteins. *Nature* 144:71–72.
15. Chibnall, A.C., Rees, M.W., Williams, E.F. and Boyland, E. (1940) Glutamic acid of proteins. *Nature* 145:311–312.
16. Chibnall, A.C. (1940) The glutamic acid of normal and malignant tissue proteins. *Biochem J* 34:285–300.
17. Dittmar, C. (1940) Über das Vorkommen von *d*(–)-Glutaminsäure in Tumorproteinen. I and II. *Z Krebsforsch* 49:397–398 and 441–442.
18. Dittmar, C. (1940) Über das Vorkommen von *d*(–)-Glutaminsäure in Tumorproteinen. III. Mitteilung. *Z Krebsforsch* 50:472–477.
19. Graff, S. (1940) Glutamic acid in malignant tumors. *J Biol Chem* 130:13–17.
20. Graff, S., Rittenberg, D. and Foster, G.L. (1940) The glutamic acid of malignant tumors. *J Biol Chem* 133:745–752.
21. Johnson, J.M. (1940) A comparison of the optical forms of glutamic acid from normal and cancerous tissues. *J Biol Chem* 132:781–782.
22. Klingmüller, V. (1943) Glutaminsäure aus Tumoren. *Z Physiol Chem* 278:97–119.
23. Konikova, A. (1940) Glutamic acid of proteins. *Nature* 145:312.
24. Town, B.W. (1940) Glutamic acid of proteins. *Nature* 145:312–313.
25. Wieland, T. (1942) Über die Glutaminsäure aus Tumoren. *Ber Deut Chem Ges* 75B: 1001–1007.
26. Wieland, T. and Paul, W. (1944) Bestimmung von *l*- und *d*-Glutaminsäure in Hydrolysate von Brown-Pearce Tumoren mit ^{15}N-Glutaminsäure. *Ber Deut Chem Ges* 77:34–44.
27. Woodward, G.E., Reinhart, F.E. and Dohan, J.S. (1941) The glutamic acid of malignant tissue proteins. *J Biol Chem* 138:677–688.
28. Kögl, F., Erxleben, H. and Herken, H. (1940) Weitere Beiträge zur stereochemischen Analyse von Tumorproteinen. *Z Physiol Chem* 263:107–124.
29. Zhao, M., Bada, J.L. and Ahern, T.J. (1989) Racemization rates of asparagine-aspartic acid residues in lysozyme at 100°C as a function of pH. *Bioorg Chem* 17:36–40.

30. Arnow, L.E., Opsahl, J.C. (1940) Partial racemization of glutamic acid in boiling hydrochloric acid solutions. *J Biol Chem* 133:765–766.
31. Johnson, J.M. (1940) The racemization of glutamic acid. *J Biol Chem* 134:459.
32. Lipmann, F., Behrens, O.K., Kabit, E.A. and Burk, D. (1940) The determination of the total D-amino acid content of human tumors and normal tissues by means of D-amino acid oxidase. *Science* 91:21–23.
33. Kögl, F., Herken, H. and Erxleben, H. (1940) Über die Sonderstellung der *d*-Glutaminsäure beim Versuchen mit *d*-Aminosäure. *Z Physiol Chem* 264:220–239.
34. Berezov, T.T. (1968) Deamination of L- and D-isomers of dicarboxylic acids in malignant tumors. *Tr Univ Druzhby Nar* 31:81–88.
35. Boulanger, P. and Osteux, R. (1949) Caracterisation et dosage de l'acide d-glutamique dans les proteines de tissus normaux et neoplastiques et les proteines microbiennes. *Compt Rend* 229:311–313.
36. Ratner, S. (1944) Conversion of *d*-glutamic acid to pyrrolidone carboxylic acid by the rat. *J Biol Chem* 152:559–564.
37. Kögl, F., Barendergt, T.J. and Klein, A.J. (1948) Excretion of D-pyrrolidone carboxylic acid by rat and dog fed with tumor proteins. *Nature* 162:732–733.
38. Hillmann, G., Hillmann-Elies, A. and Methfessel, F. (1954) Metabolism and the occurrence of d(–)-glutamic acid in tumour proteins. *Nature* 4426:403–404.
39. Mayer, C. (1951) Relation between cancer and D-glutamic acid. *Paris Méd* 41:155–156.
40. Smith, G.G. and Wonnacott, D.M. (1980) A convenient, rapid method for the resolution of enantiomeric amino acids using chiral phases on stainless-steel capillary gas chromatographic columns. *Anal Biochem* 109:414–420.
41. Stroud, E.D. and Smith, G.G. (1984) A search for D-amino acids in tumor tissue. *Biochem Med* 31:254–256.
42. Fisher, G.F., Torres, D., Bruna, J., Bergljung, C., Cerwinski, S., Martin, T., Gruneiro, A., Chou, S.-J., Man, E.H. and Pappatheodorou, S. (1995) Presence of D-Aspartate and D-Glutamate in Tumor Proteins. *Cancer Biochem Biophys* 15:79–82.
43. Aswad, D.W. (1984) Determination of D- and L-aspartate in amino acid mixtures by high-performance liquid chromatography after derivatization with a chiral adduct of *o*-phthaldialdehyde. *Anal Biochem* 137:405–407.
44. Sun, A.S. and Cederbaum, A.I. (1980) Oxidoreductase activities in normal rat liver, tumor-bearing rat liver and hepatoma HC-252. *Cancer Res* 40:4677–4681.
45. Abellatif, A.G., Preat, V., Vamecq, J., Nilsson, R. and Roberfroid, M. (1990) Peroxisome proliferation and modulation of rat liver carcinogenesis by 2,4-dichlorophenoxyacetic acid, 2,4,5-trichlorophenoxyacetic acid, perfluorooctanoic acid and nafenopin. *Carcinogenesis* 11:1899–902.
46. Tamemasa, O., Goto, R. and Suzuki, T. (1978) Preferential incorporation of some ^{14}C-labeled D-amino acids into tumor-bearing animals. *Gann* 69:517–523.
47. Goto, R., Suzuki, T. and Tamemasa, O. (1979) Characteristics of D-leucine uptake by mouse Ehrlich ascites tumor cells. *J Biochem Tokyo* 86:363–369.
48. Goto, R., Jisajima, M. and Tamemasa, O. (1980) Characteristics of D- and L-alanine transport into mouse Ehrlich ascites tumor cells. *Chem Pharm Bull Tokyo* 28:1200–1206.
49. Tamemasa, O., Goto, R., Takeda, A., Maruo, K. (1982) High uptake of ^{14}C-labeled D-amino acids by various tumors. *Gann* 73:147–153.
50. Takeda, A., Goto, R., Tamemasa, O., Chaney, J.E. and Digenis, G.A. (1984) Biological evaluation of radiolabeled D-methionine as a parent compound in potential nuclear imaging. *Radioisotopes* 33:213–217.
51. Schober, O., Duden, C., Meyer, G.-J., Müller, J.A. and Hundeshagen, H. (1987) Nonselective transport of [^{11}C-methyl]-L- and D-methionine into a malignant glioma. *Eur J Nucl Med* 13:103–105.
52. Man, E.H. and Bada, J.L. (1987) Dietary D-amino acids. *Annu Rev Nutr* 7:209–225.
53. Friedman, M. (1991) Formation, nutritional value, and safety of D-amino acids. In: M. Friedman (ed.): *Nutritional and Toxicological Consequences of Food Processing*. Plenum Press, New York, 447–480.

D-Amino Acids in Sequences of Secreted Peptides of Multicellular Organisms
ed. by P. Jollès

D-Amino acids in aging erythrocytes

D. Ingrosso and A.F. Perna

Institute of Biochemistry of Macromolecules, S.U.N. School of Medicine and Surgery, Via Constantinopoli 16, I-80138 Naples, Italy

Summary. Mature human erythrocytes are highly differentiated cells which have lost the ability to biosynthesize proteins *de novo*. During cell aging in circulation, erythrocyte proteins undergo spontaneous postbiosynthetic modifications, regarded as "protein fatigue" damage, which include formation of isomerized and/or racemized aspartyl residues. These damaged proteins cannot be replaced by new molecules; nevertheless, data support the notion that they can be repaired to a significant extent, through an enzymatic transmethylation reaction. This repair reaction has therefore been used as a means to monitor the increase of altered aspartyl residues in erythrocyte membrane proteins during cell aging. The relationship between protein repair and aspartyl racemization in red blood cell stress and disease is discussed.

Introduction

Although L-amino acids are by far more predominant in living organisms, in has now been firmly established that D-amino acids are present as natural compounds in bacteria, invertebrates and mammals. Several human tissues (endocrine and nervous) are endowed with substantial quantities of free D-amino acids, with a possible role in the modulation of hormone synthesis and release, and neurotransmission.

D-amino acid residues, that is the presence of D-amino acids within protein sequences, and their possible functions, have been well studied in erythrocyte membrane proteins and are the focus of the present chapter. We will address two basic questions: (i) the nature of the residue(s) involved, as well as the mechanism(s) for racemization, and (ii) how these alterations are related to red blood cell aging.

Regarding the first question, we will restrict our discussion to the mechanism(s) involved in the occurrence of D-aspartyl residues, by all accounts the predominant racemized amino acid species in erythrocyte proteins. The mechanisms through which D-aspartyl sites are the product of posttranslational deamidation of intrinsically labile L-asparaginyl residues, and/or isomerization of L-aspartyl residues, will be reviewed.

A central issue is the versatility of the erythrocyte as a convenient model to study this postbiosynthetic reaction. In fact, in these cells, protein *de novo* biosynthesis is nonexistent because of the lack of the ribosomal and nuclear machinery necessary for it. The aging of these cells in circulation is paralleled by the occurrence of a number of spontaneous modifications of proteins. Since these "aged" proteins cannot be replaced by new molecules, they are expected to last as long as the cell itself, unless they are either removed or repaired.

The "protein molecular fatigue" concept

Spontaneous postbiosynthetic modifications are widespread reactions, including deamidation of asparaginyl and glutamyl residues, racemization of aspartyl residues, glycation of α- and ε- free amino groups, oxidation of methionyl, cysteinyl and triptophanyl residues [1, 2]. They are generally considered to be deleterious to protein structure and function, although qualified exceptions are known, notably the binding of retinal to opsin in the mechanism of vision, for example [2].

Most of these spontaneous events can be characterized as degradative reactions, affecting the structure and function of these macromolecules, during the course of their aging. The propensity of protein structures to undergo spontaneous molecular modifications has been recently referred to as "protein fatigue" [3]. The rates of these reactions depend on specific sequence and three-dimensional structure determinants, and are influenced by the microenvironment.

As a general principle, protein fatigue damage, involving even a few residues within a protein, can cause significant functional impairment. This is not surprising, considering that a single amino acid substitution, as a consequence of a point mutation in the relevant genes, has the potential to profoundly affect both the structural and functional properties of certain proteins. In this respect, the effects of deamidation on protein charge, or the ability of racemization to destabilize secondary structure, are straightforward consequences. Also, the longer the protein lifetime, the higher the probability of these fatigue events. The outcomes, in terms of loss of function, are predictably crucial in terminally differentiated cells, where protein *de novo* biosynthesis can be deactivated. In these tissues, and particularly in long-lived proteins, the stochastic accumulation of molecular fatigue damage reaches detectable levels, which can significantly affect protein structure and function [3].

Protein fatigue should not be seen solely as the unavoidable consequence of aging of these molecules; as we will see further in the chapter, it can be also influenced by particular cell stress conditions. An analogy can be made to the phenomenon of metal fatigue in aircraft, where the effects of repeated stress cycles can lead to critical damage of metal structures.

The origin of isomerized and racemized aspartyl residues in proteins

Only L-amino acids are used for protein biosynthesis in living cells. Nevertheless, it is generally accepted [2] that L-amino acid residues in proteins can slowly racemize, through the formation of a planar carbanion, yielding the corresponding enantiomers:

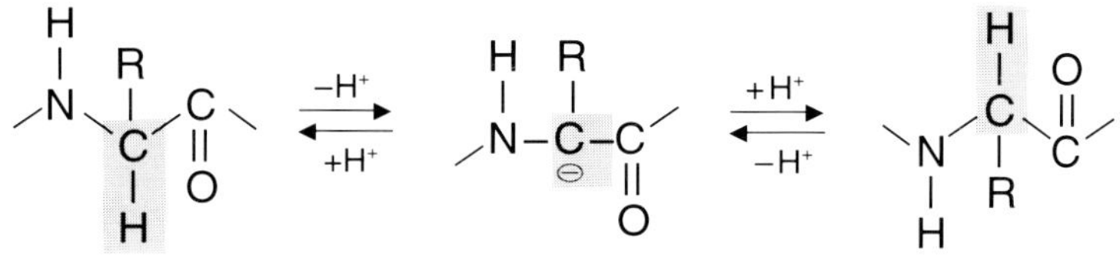

For practical purposes, since the rate of accumulation of such racemized residues would be very slow, compared to the lifetime of most proteins, D-amino acids should be expected to be almost undetectable. Moreover, the relatively high background due to racemization occurring during protein treatment under strong acidic conditions, necessary to achieve complete hydrolysis, can make quantitation of naturally occurring D-amino acid residues in proteins rather difficult to accomplish.

On the other hand, it has been shown that, while the peptide bond is rather stable, the reactivity of the side chains of some residues significantly contributes to general protein instability. Carboxamide residues are particularly unstable, in that they are prone to deamidation [2–5]. Cyclization of amino terminal glutamine is a frequent reaction occurring in proteins, both *in vivo* and during purification and characterization procedures, thus leading to pyroglutamate formation. Also, asparagine is one of the most labile residues, due to its susceptibility to deamidation. Because of the time dependency of this phenomenon, it has been proposed that, through its deamidation, asparagine could represent a potential built-in timer, a sort of indelible marker of aging of partially degraded and functionally impaired protein molecules [6].

Deamidation of intrinsically labile L-asparaginyl residues, through a cyclic succinimide intermediate, represents the major source of L-isoaspartyl residues in proteins [7–10]. Details of this mechanism, which has been clarified by independent investigators [3, 5, 11, 12], are given in Figure 1. In the L-isoaspartyl residue, the regular alternation of nitrogen and carbon atoms in the peptide backbone is interrupted by the presence of an extra methylene group, since the aspartyl is linked through its β-carbonyl to the subsequent residue in the peptide chain, which can result in the destabilization of local protein conformation.

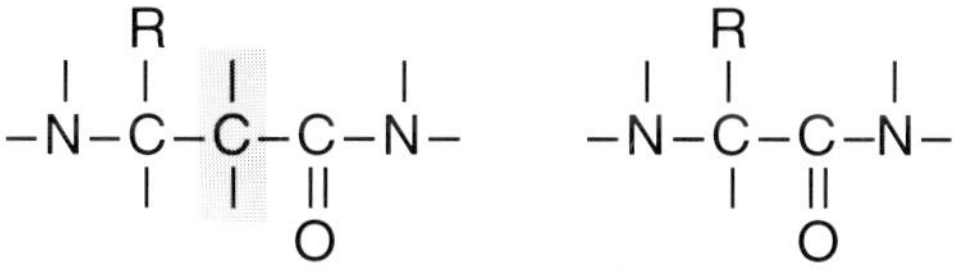

L-isoAsp L-Asp

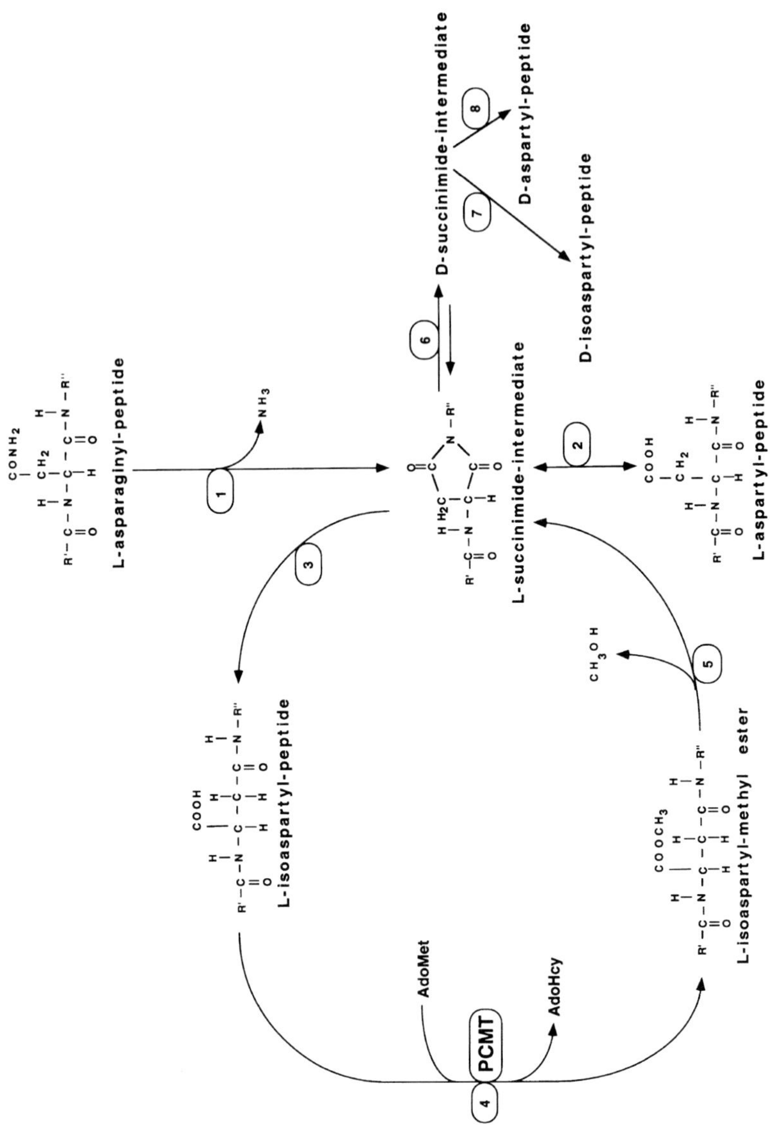

Figure 1. Mechanism for spontaneous degradation and repair of asparaginyl and aspartyl residues in proteins. The α-nitrogen of the Asn^{+1} residue attacks the β-carbonyl carbon of the Asn residue, with the loss of ammonia (step 1). The thus formed succinimidyl ring can open on either side of the imide nitrogen, generating an L-aspartyl residue (step 2), or an L-isoaspartyl residue (step 3), respectively. Data obtained on different model peptides and proteins have shown that the L-isoaspartyl form is always favored over the L-aspartyl form, with a 7:3 ratio, respectively. The L-succinimide can slowly and reversibly racemize (step 6), yielding D-isoaspartyl- (step 7) and D-aspartyl- (step 8) containing peptides respectively. The figure also describes the repair mechanism of the isopeptide bond at the level of altered L-isoaspartyl residues. This consists of two reactions: the first (step 4) is the enzymatic methyl esterification of the free α-carboxyl group, catalyzed by protein L-isoaspartyl (D-aspartyl) *O*-methyltransferase (PCMT; EC 2.1.1.77); the second (step 5) is the spontaneous methyl ester hydrolysis, leading again to L-succinimide. Repeated cycles of methylation and demethylation allow the quantitative conversion of L-isoaspartyl residues into normal L-aspartyl residues. The repair mechanism also predicts the formation of racemized residues (steps 6 through 8), as side products. AdoMet: *S*-adenosylmethionine. AdoHcy: *S*-adenosylhomocysteine.

The effects of a number of structural elements, as well as of the protein microenvironment, have to be taken into account in trying to predict the propensity of a given Asn residue to undergo deamidation. It has been demonstrated that the nature of the Asn^{+1} residue is crucial. The most labile asparagine residues are the ones which are followed by small nonbulky side-chain residues, such as Gly, Ser and Ala [7, 9, 13–15]. From studies on model peptides, it has been calculated that the rate of succinimide formation reaches its maximum in case of Asn-Gly sequences, which are able to form the cyclic derivative 6.5–17.6 times faster than asparagine followed by alanine [16]. Moreover, succinimide formation occurs at an intermediate rate, with values between those of Asn-Gly and of Asn-Ala, when asparagine is flanked by a serine residue [16]. The majority of asparagine deamidation events in proteins is consistently associated with Asn-Gly and Asn-Ser sequences, although the occurrence of alternative residues at the Asn^{+1} position, in a number of naturally deamidated proteins, could be explained by the existence of different deamidation mechanisms [13]. The analysis of the nearest-neighbor frequency for asparagine residues, carried out on a database including over a thousand proteins, indicated that nonpolar bulky residues are favored compared with smaller and more reactive ones. This would suggest that deamidation, as a spontaneous chemical accident affecting protein structures, was selected against during the course of evolution [13, 14].

The role of three-dimensional organization and/or of ligand binding on protein susceptibility to this specific fatigue event has been explored using different protein and peptide models *in vitro*. The calcium-binding protein calmodulin has been a very useful model for these studies. Altered aspartyl residues have been found to occur in this protein both *in vitro* and *in vivo*, in intact cells [3, 17–20]. However, deamidation of susceptible sequences in calcium-binding domains can be prevented, *in vitro*, when calcium is bound to the molecule, due to the loss of flexibility of the relevant regions [13, 21].

The role of conformational flexibility as a major determinant in asparagine deamidation is further supported by the fact that, in mouse epidermal growth factor, deamidation of residue Asn^{16} is triggered by protein unfolding [22]. Conversely, it has been shown that deamidation of Asn^{67}, yielding an L-isoaspartyl residue, occurs spontaneously *in vivo* in bovine seminal ribonuclease, although this residue is constricted within a rather rigid loop, locked by a disulfide bridge [10, 23]. Rabbit liver hydroxymethyltransferase [24] and human triosephosphate isomerase [25] are further examples of proteins undergoing deamidation *in vivo*.

A crucial point regards the reflections of these structural alterations on protein function. It has been shown, in this respect, than asparagine deamidation is able to affect the biological activity of certain proteins, including human growth hormone-releasing factor [26], human epidermal growth factor [27], lysozyme [28], hemoglobin Providence [29]. In addition, *in vitro* deamidation of calmodulin yields two derivatives characterized as an aspartyl form and an isoaspartyl form, which respectively show only 40 and 10% of the native protein biological activity [18].

Succinimide formation from a normal L-aspartyl residue (Fig. 1) has been described to occur in model peptides of known sequence, although at a rate 13.1–35.6 times slower than in the case of the corresponding L-asparaginyl peptides [16]. Such unstable L-aspartyl residues occur in proteins such as glucagon [30] and human growth hormone [31]. Interestingly, the rate of L-aspartyl isomerization, unlike that of asparaginyl deamidation, is barely modified in conditions where the peptide is solubilized in organic cosolvent. This may indicate that the asparaginyl residues in contact with protein internal regions of the membrane-spanning domain are relatively stable, while the stability of L-aspartyl residues, in the same conditions, is poorly influenced by changes in the dielectric constant of the microenvironment [32–33].

It has been hypothesized that amino acid racemization could be involved in the molecular mechanisms of diseases. Particularly interesting are the potential implications of D-amino acid residues in the pathophysiology of Alzheimer's disease. The β-amyloid protein extracted from the brains of patients affected by this disease has been shown to contain high levels of D-aspartyl and D-isoaspartyl residues [34–36]. It has been reported that the presence of such residues can be responsible for a significant reduction in β-amyloid solubility, and this not only *in vitro* [34, 37], but it may also influence the accumulation in brain of this key protein during disease development. Other long-lived proteins where high levels of D-aspartyl residues occur are the dentine in teeth [38] and the crystallins of eye lens [39]. These promising studies, still in progress, shed new light on our understanding of the possible role of D-aspartyl residue formation in the progression of diseases such as lens cataracts and Alzheimer's disease.

Racemization of amino acid residues in proteins is commonly referred to as a very slowly occurring event, so it has been proposed as a suitable age marker not only in living tissues but also in biological remains, and has in fact been used to estimate the time since death of biological samples [40]. Erythrocyte proteins, particularly membrane components, represent a remarkable exception to this general interpretation, since their racemization rate, specifically for what relates to aspartyl residues, is unexpectedly higher than that of other model proteins [41, 42]. It has been calculated that the average *t* 1% (time required to convert 1% of the L-Asx residues into their D form, including D-aspartyl and D-isoaspartyl), is, in the case of erythrocyte membrane proteins, 38.6 days. This time is not equal for all membrane proteins, since the intrinsic ones are slightly less stable than the extrinsic ones [41], but in general the propensity to deamidation of membrane-related proteins is, in red blood cells, significantly higher (about four times), than cytosolic proteins [41]. It is worth noting that racemization rates of membrane proteins are fully comparable with the 120-day duration of the erythrocyte life span in humans, that is these macromolecules have higher probability than other proteins to actually accumulate racemized aspartyl residues during the course of their aging in circulation. The potential implications of this phenomenon, in terms of protein aging and repair, will be discussed in the next sections.

L-Isoaspartyl (D-aspartyl) *O*-methyltransferase, a repair enzyme for fatigue-damaged proteins

Covalent enzymatic modifications are in most cases acknowledged regulatory reactions, which reversibly or irreversibly modify protein structure and function [1]. This is also true for some of the *S*-adenosylmethione-dependent methyltransferases, which recognize specific proteins as methyl-accepting substrates [3, 5]. However, during the last decade, experimental evidence showed the existence of a ubiquitous enzyme, the protein L-isoaspartyl (D-aspartyl) *O*-methyltransferase (PCMT; EC 2.1.1.77), endowed with a very peculiar site-specificity toward fatigue-damaged peptides and proteins [3, 5].

Studies on the substrate specificity of PCMT have been performed using either deamidated protein or peptides of known sequence containing altered aspartyl residues. The results showed that the major methyl-accepting site is the free α-carboxyl group of the L-isoaspartyl residues, whose K_m values are in the micromolar range [3, 5]. Among the other tested peptides, D-aspartyl residues have been found to be very poorly recognized *in vitro* because of a K_m value situated in the millimolar range, while D-isoaspartyl residues are not modified at all by PCMT [43, 44]. The potential meaning of D-aspartyl methyl esterification, *in vivo*, will be further discussed in the next paragraph. Nevertheless, these *in vitro* studies have provided the basis of enzyme-substrate interaction models [43, 45].

PCMT is a constitutive protein isolated and characterized from several mammalian sources. The human enzyme shows wide tissue distribution. In erythrocytes, as well as in other cells, two major isoenzymes, of 226 and 227 residues respectively, have been structurally characterized, and differ in their isoelectric points by one pH unit [46, 47]. A minor isoenzyme form has been also identified in erythrocytes, with an isoelectric point intermediate between the major forms [48]. The two major isozymes display the same substrate specificity, at least *in vitro* [49], and are generated through an alternative splicing mechanism of the single gene transcript, including eight exons, interrupted by seven introns, located on chromosome 6 [47, 49–51]. Structural studies also provided more recent insights on the relationships between genetic polymorphism and thermal stability of PCMT, as well as on the existence of common structural motifs of potential functional meaning in a vast group of *S*-adenosylmethionine-dependent methyltransferases [46, 52, 53].

PCMT has been widely used as a tool for *in vitro* identification of L-isoaspartyl residues in natural and recombinant proteins [3, 5, 54–57]. The search for the actual function of an enzyme with such a peculiar substrate specificity has kept scientists striving for more than a decade. It has been shown, by several research groups independently, that *in vitro* incubation of L-isoaspartyl-containing peptides with purified PCMT and the methyl donor *S*-adenosylmethionine leads to quantitative conversion of the L-iso-aspartyl peptide into the normal L-aspartyl derivative [3, 5]. For example, this was the case of [iso-Asp^3]tetragastrin [58], of isopeptides derived from bovine seminal ribonuclease [23], as well as from both lactate dehydrogenase and adenocorticotropin [59], and of *Escherichia coli* HPr phosphocarrier protein [60].

According to the proposed repair mechanism, depicted in Figure 1, methyl esterification activates the free α-carboxyl group of the isoaspartyl residue. The methyl ester itself is unstable, and spontaneously hydrolyzes through the formation of the same L-succinimide derivative that results from asparaginyl deamidation. From the opening of the ring, about 30% recovery of the normal L-aspartyl derivative can be predicted to occur at each methylation-demethylation cycle. The ability of this combined enzymatic-spontaneous pathway to restore the biological function of proteins inactivated by deamidation has been also confirmed. For example, enzymatic methyl esterification is able to restore *in vitro* a significant portion – about 50% – of the biological activity of deamidated calmodulin [18]. It is worth noting that calmodulin is also methyl-esterified in intact erythrocytes [19].

It should be underscored that, because of the propensity of L-succinimidyl residues to racemize, and since D-aspartyl residues are relatively poor substrates for PCMT, the repair mechanism predicts that racemized residues, particularly D-isoaspartyl residues, will slowly accumulate as by products of L-isoaspartyl residue repair [44, 61]. In fact, mathematical modeling of the spontaneous aging and enzymatic methylation reactions in erythrocytes shows that PCMT could be responsible for a significant increase in the number of racemized aspartyl residues during the 120-day life span of these cells in circulation [62]. This model is supported by three lines of evidence: (i) enzymatic methylation of an L-isoaspartyl-containing form of *E. coli* HPr phosphocarrier protein leads to *in vitro* conversion of L-isoaspartyl residues into L-aspartyl residues, with the generation of a small but significant number of D-isoaspartyl residues [60]; (ii) D-isoAsp buildup increases in parallel with repair of other L-isoaspartyl-containing peptides *in vitro* [58, 59, 63]; (iii) human α A-crystallin purified from eye lens of elderly individuals contains primarily D-isoaspartyl and L-aspartyl residues at positions 58 and 151, suggesting that the active PCMT in this tissue repaired L-isoaspartyl residues *in vivo* [64].

According to this model, the repair of the major forms of asparagine and aspartate damage (that is L-isoaspartyl and D-aspartyl, respectively) can paradoxically result in the accumulation of an additional altered form, the D-isoaspartyl residue. On a quantitative basis one should consider that the occurrence of such amounts of racemized aspartyl residues in proteins appears as a relatively low price to pay in exchange of having proteins virtually free of L-isoaspartyl residues [62]. In this respect, studies of molecular models concluded that the position of the D-isoaspartyl side-chain carboxyl group in space is in fact similar to that of the L-aspartyl residue, and different from that of both L-isoaspartyl and D-aspartyl residues [9]. The appearance of such D-isoaspartyl sites may therefore be less detrimental to protein structure and function than either D-aspartyl or L-isoaspartyl sites [44].

PCMT itself is subject to molecular-fatigue damage. Automethylation of the bovine erythrocyte enzyme has been actually shown to occur *in vitro* in at least two sequence positions (i. e., Asn^{188} and Asp^{217}), raising the possibility that self-repair of the enzyme could actively contribute to the maintenance of repair system efficiency [65, 66].

Evidence has been obtained that PCMT is also able to recognize damaged peptides and proteins *in vivo*. It has been shown that when HeLa cells are subject to heat shock, an increasing rate of methyl esterification of the endogenous substrates is detectable. This is due either to an augmented formation of, or to improved accessibility to, the methyl-accepting sites in heat shock-damaged protein substrates. No PCMT induction is observable under the same conditions [67]. The biological function of PCMT in protein repair *in vivo* is also supported by studies on *Xenopus laevis* oocytes [68], where endogenous calmodulin is methyl-esterified to an extent comparable to that of the erythrocyte calmodulin [20]. In addition, it has been shown that L-isoaspartyl-containing peptides, microinjected into oocytes, are selectively recognized by PCMT [69].

Consistent with these results are the data obtained with the nematode *Caenorhabditis elegans*, which indicate that PCMT plays a crucial role in the *dauer* larval form, a developmental stage with the potential for long-term survival. Moreover, the fact that the nematode enzyme does not recognize D-aspartyl residues would indicate that this substrate specificity was a late evolutionary adaptation [70].

In *Drosophila melanogaster*, PCMT gene transcript increases in parallel with enzyme activity during development. The fact that PCMT activity levels still remain constitutively high in adult organisms supports its role during aging [71].

An elegant demonstration of the role of PCMT in the repair of altered aspartyl residues in proteins is provided by recent results using PCMT gene-targeting technology in mouse embryonic stem cells [72]. The relevant methyltransferase knockout mice die prematurely at an average age of 42 days after birth. Abnormally high levels of protein substrates for PCMT (i. e., damaged proteins, containing altered aspartyl residues) could be evidenced in the cytosolic cell fraction of several tissues and organs, including liver, brain, heart and erythrocytes.

The evidence so far reported speaks in favor of the repair role of PCMT in intracellular proteins, whereas extracellular proteins, though also expected to undergo molecular fatigue, would simply not be accessible for the repair enzyme, which is confined within the intracellular space. Nevertheless, it has recently been reported that PCMT actually becomes detectable in extracellular space, at the nearby site of blood vessel injury [73]. In addition, it has been shown that major substrates of this injury-released PCMT are proteins of the extracellular matrix, and, in particular, type I and type III collagen [74].

This, as well as other evidence [3, 5], points to the role of PCMT in preventing the accumulation of altered aspartyl residues which spontaneously arise in aging protein molecules *in vivo*.

Enzymatic methylation of damaged aspartyl residues in erythrocyte proteins

Influence of cell aging

The human erythrocyte is a highly differentiated cell, which has lost the nucleus, as well as all other intracellular organelles. The only membrane system still present is the plasma membrane, whose protein components are fairly well characterized [75]. A membrane cytoskeleton, made of several different protein components, is anchored to the lipid bilayer by means of specific proteins (Fig. 2). The membrane-cytosketon complex is thus responsible for the maintenance of red blood cell shape, plasticity and deformability, particularly during its passage through the small capillaries. Loss of the ability to smoothly advance with the blood flow, by adapting to the diameter of the narrow vessels, can in fact be responsible for early removal of erythrocytes from circulation, a phenomenon which, under normal physiological conditions, is expected to occur at spleen level after about 120 days of permanence within circulation [76].

Erythrocyte aging is nevertheless accompanied by a number of progressive alterations of its components, including several cytoplasmic enzymes, whose decay of activity can be used as a marker of erythrocyte aging [77]. Due to their availability, human erythrocytes have also been used by investigators interested in the relationships between posttranslational modification of proteins, both spontaneous and enzymatic, and cell aging [76]. In this respect red blood cells represent an ideal model system, since protein biosynthesis is blocked by the lack of ribosomal and nuclear machinery necessary for it. Therefore, we expect that protein fatigue damage would accumulate in erythrocyte proteins, while these molecules connot be replaced by newly synthesized macromolecules as the red blood cells age in circulation [3].

Based on the information provided in the previous paragraph, the *link* between the formation of altered aspartyl residues in proteins and red blood cell aging could indeed be the protein methyl esterification reaction

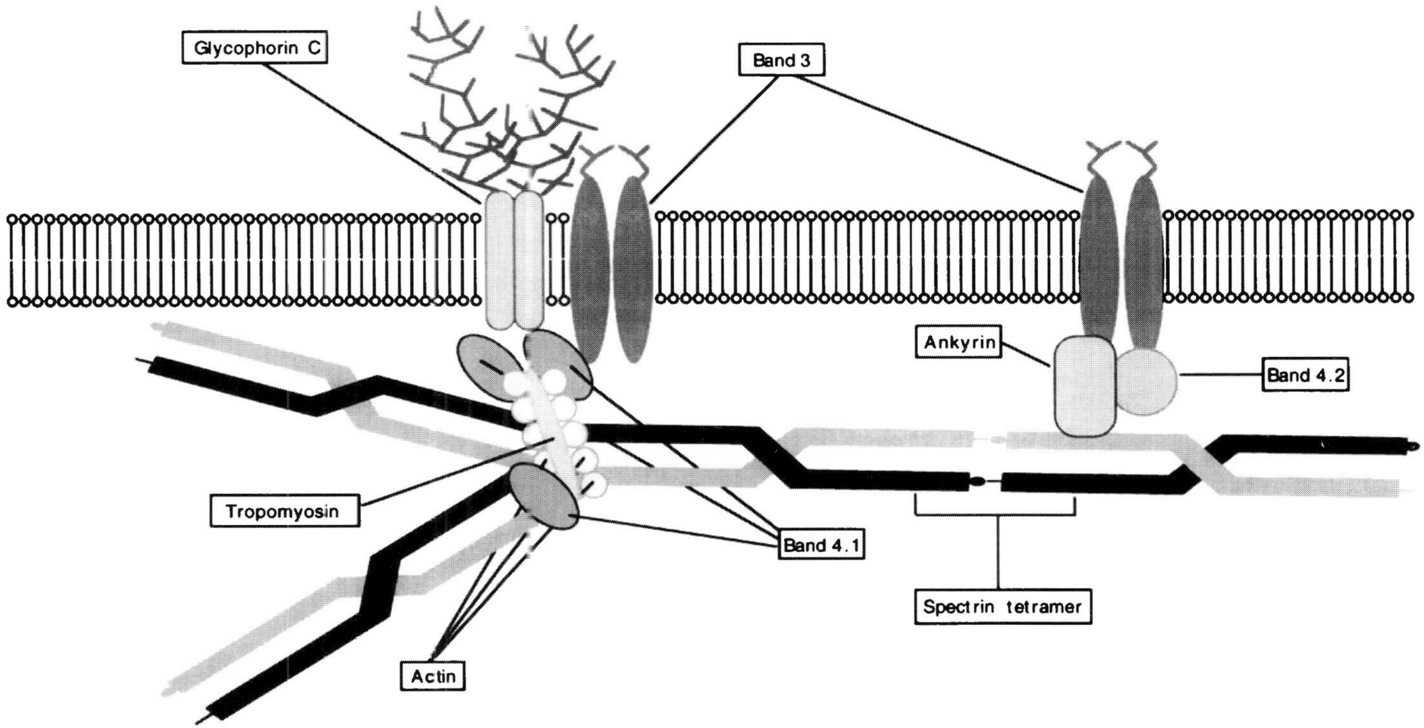

Figure 2. Schematic representation of human erythrocyte membrane proteins. The principal protein components are indicated according to a common nomenclature. The dimeric intrinsic protein band 3 is the membrane anion carrier (AE1), as well as a crucial site of attachment of the cytoskeleton to the membrane. The cytoskeleton is a protein network whose major components include α-spectrin and β-spectrin, organized in tetramers, which form, together with actin, a junctional complex, for whose stability protein 4.1 plays a critical role. Extrinsic proteins, such as ankyrin and the actin-binding protein band 4.1, mediate the anchorage of the cytoskeleton to the above membrane. The intracellular domain of band 3 also interacts with band 4.2 The whole membrane-cytoskeleton complex is stabilized through these mutual intermolecular interactions, which also ensure the physico-chemical properties of the erythrocyte membrane in terms of maintenance of discoid shape, deformability and plasticity. The major methyl-accepting proteins in intact erythrocytes are band 3, as well as the cytoskeletal-related components ankyrin, band 4.1 and band 4.2.

catalyzed by PCMT. In this respect, it has been demonstrated that the erythrocyte possesses the metabolic machinery for adenosylmethionine biosynthesis as well as for the removal of its demethylated product adenosylhomocysteine (Fig. 3). This appears to be very important for healthy red blood cell metabolism, since this thioester is a powerful inhibitor of all adenosylmethione-dependent methyltransferases, including PCMT [78, 79]. In addition, in these cells the PCMT reaction is by far the major adenosylmethionine-consuming reaction in the erythrocytes [80]. Although PCMT is a cytosolic enzyme, many of its major *in vivo* substrates are in fact membrane proteins, including integral membrane protein band 3, as well as cytoskeletal components ankyrin, band 4.1 [81] and 4.2 [79].

Cytosolic proteins, including hemoglobin and carbonic anhydrase [82], have also been shown to be methylated, although to a generally low extent. Interestingly, it has been reported that, consistent with previous findings using other model proteins [23], PCMT preferentially recognizes partially denatured hemoglobin molecules, with an increased tendency to bind to the

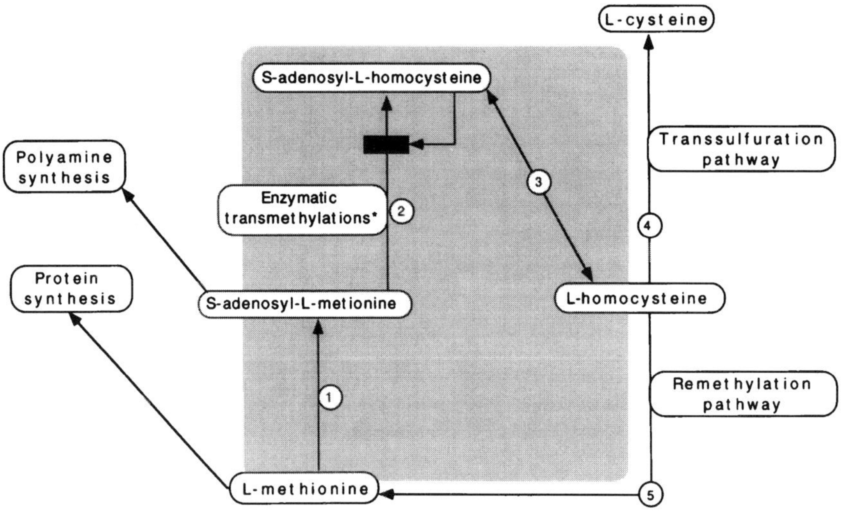

Figure 3. Schematic representation of metabolic pathways involved in enzymatic protein methylation. The gray-shaded area refers to reactions taking place in the erythrocyte. Step 1 is the biosynthesis of the methyl donor *S*-adenosylmethionine (AdoMet), from L-methionine and ATP, catalyzed by AdoMet-synthetase. Step 2 includes all AdoMet-dependent transmethylation reactions. *S*-adenosylhomocysteine (AdoHcy), the demethylated AdoMet product, is also the powerful competitive inhibitor of all AdoMet-dependent enzymatic transmethylations. Under physiological conditions such inhibition is prevented by reversible hydrolysis of AdoHcy to adenosine and L-homocysteine, catalyzed by AdoHcy-hydrolase (step 3). Prompt removal of hydrolysis products prevents reaction 3 from proceeding in the biosynthetic direction. L-homocysteine can be further metabolized in other organs (step 4 and 5), such as the liver, and/or excreted by the kidney predominantly as its oxidation product. An asterisk indicates that the only AdoMet-dependent reaction quantitatively relevant to the erythrocyte is the one catalyzed by L-isoaspartyl (D-aspartyl) *O*-methyltransferase. Intraerythrocyte adenosylhomocysteine accumulation, due to hyperhomocysteinemia, has been found to occur in uremia [79, 93–95].

membrane, as the ones which occur during erythrocyte senescence [83]. PCMT-dependent methyl esterification of both α and β chains has been detected, but identification of methyl-accepting sites as Asp^6 (or Asn^9) has been achieved only in the case of the α chain [84]. A PCMT-independent process of hemoglobin methylation has also been described to occur, in both α and β chains, at sites different from Asx and mostly involving lysine residues; it is therefore totally distinct from the repair reaction [85].

Erythrocyte calmodulin is enzymatically methylated by PCMT in intact cells, at level of Asp^2, $Asp^{78/80}$ and Asx residues located in calcium-binding domains III and IV [19]. These results suggest that (i) calmodulin actually becomes deamidated *in vivo* due to the lack of protective effect exerted by calcium, which is present only at very low concentrations intracellularly; and (ii) PCMT-driven repair of damaged aspartyl residues in calmodulin can effectively take place *in vivo*.

In general, since L-isoaspartyl residues cannot be quantitated in proteins following acid hydrolysis, which converts them to normal L-aspartate, and specific sites prone to damage have not as yet been extensively identified in red blood cell proteins, methylation by PCMT is the best method available for identifying L-isoaspartyl residues. Measurements of methylation levels of endogenous proteins can therefore be used to monitor the occurrence of underlying spontaneous molecular fatigue damage, provided that erythrocytes are still metabolically intact.

According to these concepts, by using different age fractions of erythrocytes, obtained through isopicnic centrifugation, it was possible to demonstrate that methylation of endogenous membrane proteins significantly increases as cells age in circulation (Fig. 4) [86, 87]. This increase was particularly dramatic for cytoskeletal protein components [86]. The ability of cell aging to modify other factors, connected with the methyl esterification process, including PCMT activity and adenosylmethionine synthesis, could be ruled out, so it was possible to conclude that in fact specific membrane and cytoskeletal proteins became better methyl acceptors during erythrocyte senescence because of their propensity to become intrinsically altered [86–88].

Characterization of methylated residues has been attempted by analyzing hydrolysates of membrane proteins, methyl-esterified in intact erythrocytes. This approach allowed a 22% recovery of total methyl groups, from erythrocyte membrane proteins ankyrin and band 4.1, as D-aspartyl-β-methyl esters [44]. Otherwise, the direct identification of modified L-isoaspartyl sites, which are still by far the major methyl-accepting residues in several models, has always been difficult. This is probably due either to the instability of L-isoaspartyl-α-methyl esters or to their poor recovery during membrane protein proteolytic treatment. Notably, deamidation of band 4.1 at the level of Asn^{478} and Asn^{502} has been established. In particular, deamidation of the latter residue is responsible for the time-dependent conversion of band 4.1b into its 4.1a derivative, which is associated with aging of red blood cells [89].

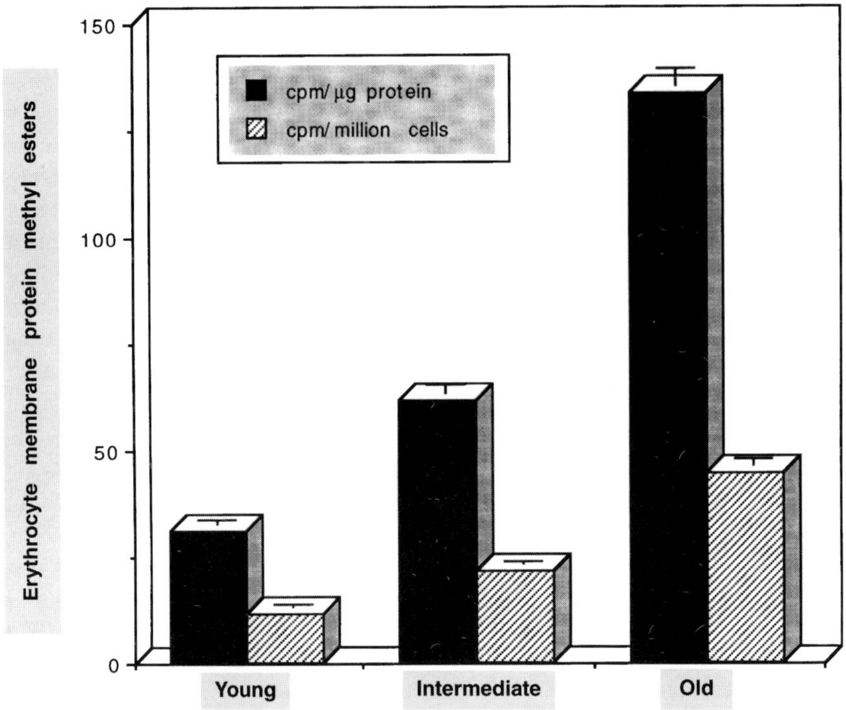

Figure 4. Enzymatic membrane protein methyl esterification increases as the erythrocyte ages in circulation. Three age-related red cell fractions were obtained by density gradient centrifugation. Individual age fractions were incubated with [methyl-^{3}H]methionine, the *in vivo* AdoMet precursor. At the end of incubation, membranes were isolated, and the radioactivity incorporated as protein methyl esters was counted. Results are expressed as both cpm/µg protein and cpm/million cells, and reported ± SE. Further details are given in ref. 86. Notably, no significant alterations in the other major metabolic pathways, gray-shaded in Figure 3, have been found to occur during erythrocyte aging [86, 87].

Implications of cell stress and disease

These reports established a clear relationship between erythrocyte aging and the increased probability of finding altered isomerized and racemized aspartyl residues in proteins, which, in turn, is opposed by the relentless repair effort undertaken by PCMT. The question is, which factors make erythrocyte membrane proteins particularly susceptible to aspartyl-related fatigue damage? Recent data show that there are in fact diseases in which the peaceful picture of the erythrocyte becoming older and older in circulation, while its PCMT promptly repairs damaged L-isoaspartyl (and D-aspartyl) residues, is disturbed.

One of these conditions is hereditary spherocytosis, a chronic haemolytic anaemia, associated with an intrinsic erythrocyte defect at the level of the membrane-cytoskeleton network, yielding a reduction of surface area/volume ratio, so that red blood cells lose their typical biconvex shape, becoming spherocytes [75]. At the molecular level this condition is associated with an isolated or combined deficiency of certain membrane-related protein components (Fig. 2), which in most cases include spectrin [75]. The result is various degrees of structural disarray of the membrane-cytoskeleton, with consequent increased osmotic fragility and reduced life span of cells in circulation.

It has been shown that spherocytosis is accompanied by an increased level of enzymatic methylation of red blood cell membrane proteins, despite the fact that the circulating red blood cell population is younger than normal controls, according to common markers of erythrocyte aging [90]. It has been proposed that this methylation increase, which correlates with the degree of spectrin deficiency, monitors a state of increased membrane protein fatigue damage, at the aspartyl residue level, due to the overall instability of the membrane-cytoskeletal network [90]. It has been also reported that the surgical removal of the spleen, a common therapeutical approach for this disease, which leads to improvement of the intracirculatory life span of spherocytes, is also associated with a significant reduction in protein methylation levels, approaching the values measured in control subjects [91]. The latter results underscore the role played by mechanical and/or metabolic stress, which red cells are subject to during their repeated passages through spleen microcirculation, thus representing a crucial trigger factor of protein fatigue damage, that is specifically monitored by increased protein methylation levels. This effect is magnified in spherocytes, where predisposing intrinsic alterations of the membrane cytoskeleton complex make cells particularly sensitive to recurring haemodynamic stress events [90, 91].

The relationship between cell stress and protein molecular fatigue is also underscored by the recent observation that membrane protein methyl esterification also rises significantly and proportionally in normal erythrocytes exposed to extensive osmotic stress [92].

A different model system is represented by chronic renal failure, a condition constantly associated with severe anemia, commonly referred to erythropoietin deficiency, but which is also accompanied by various alterations of the erythrocyte membrane, including some of its protein components [93]. In this disease, a reduction of protein methylation levels has been reported [79] which cannot be explained by a consistent reduction in red blood cell mean age (Fig. 5). This effect is associated with a significant decrease of the intracellular concentration ratio between the methyl donor, adenosylmethionine, and the methylation inhibitor, adenosylhomocysteine [79]. The presence of abnormally elevated adenosylhomocysteine is in turn due to chronic hyperhomocysteinemia [94]. The accumulation of this amino acid in plasma, which is regarded as a risk factor for cardiovascular disease, is highly prevalent in end-stage renal disease patients, and, according to the mechanism reported in Figure 3,

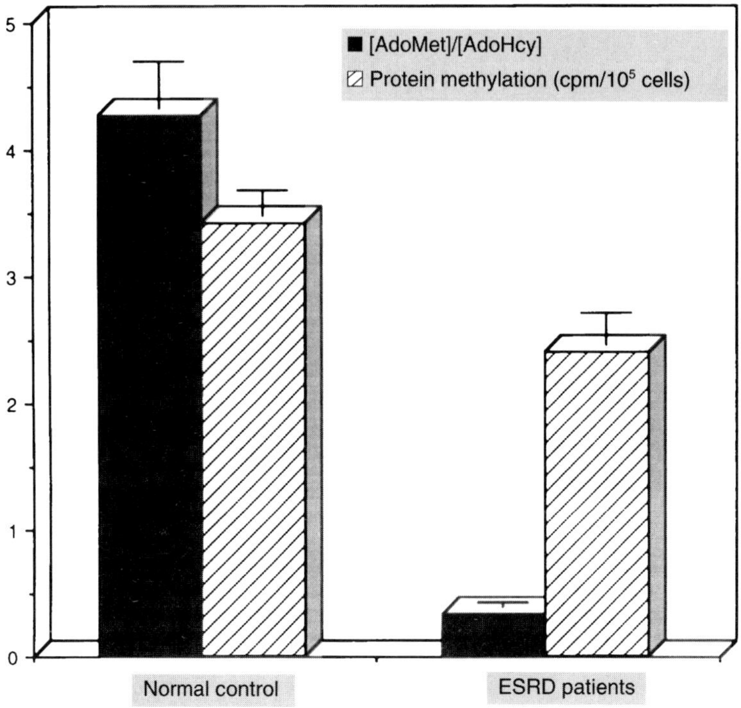

Figure 5. Inhibition of erythrocyte membrane protein methylation in uremia. Enzymatic methyl esterification of erythrocyte membrane proteins was measured in end stage renal disease (ESRD) patients and in normal controls. The results show that, in ESRD, methylation levels were significantly decreased compared with normals. In addition, a dramatic increase in the intracellular concentration of methylation inhibitor adenosylhomocysteine (AdoHcy) could also be detected in ESRD patients. This latter result is expressed as [AdoMet]/[AdoHcy] (concentration ratio adenosylmethionine/adenosylhomocysteine) ± SE. The observed differences could not be explained by consistent variations in the age of circulating red blood cells between ESRD patients and controls. Further details are given in ref. 79.

leads to high adenosylhomocysteine by interfering with adenosylhomocysteine hydrolysis reaction [93, 94]. Indeed, this disease represents an almost unique example where the PCMT-dependent protein repair system is chronically inhibited *in vivo*. The extent of this inhibition has been evaluated in erythrocyte membrane proteins from these patients by an original approach that consists in determining the extent of racemized and isomerized aspartyl residues as by products of the repair reaction [62]. Proteins were cleaved, by mild chemical and enzymatic digestions at 37°C, using specific proteases, in order to minimize the high racemization background generally associated with acidic hydrolysis procedures [62]. The results (Fig. 6) show that in normal subjects the extent of racemization of aspartyl residues in membrane proteins is in good agreement with the value of aspartyl racemization predicted by a mathematical model for an erythro-

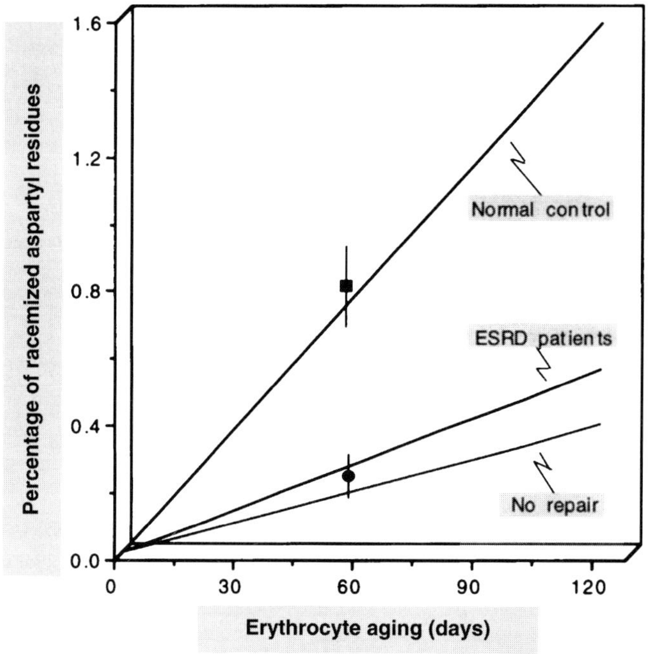

Figure 6. Extent of racemization of aspartyl residues in erythrocyte membrane proteins and its relationships with L-isoaspartyl protein repair in uremia. Racemized aspartyl residues (D-Asp and D-isoAsp) were measured by means of an enzymatic procedure which minimizes spontaneous racemization otherwise unavoidable under strong acidic conditions. Experimental data ± SE were fitted into a mathematical model simulating all reactions which generate racemized aspartyl residues, during erythrocyte aging. These would include both the spontaneous pathway [through steps 1(2)-6-7(8); Fig. 1] and the one mediated by protein repair methyltransferase [through steps 4-5-6-7(8); Fig. 1]. Straight lines depict the accumulation of D-Asp and D-isoAsp residues in normal cells (where 100% protein repair activity is present), and in the hypothesis of no protein repair activity. Experimental results showed less than 1/3 D-Asp and D-isoAsp residues in uremia (end stage renal disease, ESRD) patients compared with normals. Computer modelling predicted that such a reduction should correspond to an 11% residual activity of protein repair enzyme in erythrocytes of ESRD patients, which is in good agreement with [AdoMet]/[AdoHcy] data from Figure 5. Conversely, neither the abundance of other amino acids, nor of total D-amino acids, other than Asx, turned out to be different between patients and controls. In addition, no difference in mean age of circulating erythrocytes could account for the observed changes in D-aspartyl content between ESRD patients and control subjects. Further details are given in ref. 62.

cyte population of corresponding mean age, where PCMT is fully active [62]. Accumulation of racemized aspartyl residues in end-stage renal disease patients is about one-third that of controls, which, according to the computer simulation reported in Figure 6, corresponds to a residual PCMT activity of about 11% of normal. This value is in good agreement with the degree of derangement of the [adenosylmethionine]/[adenosylhomocysteine] ratio detected in the same disease [62, 79]. These differences in the D-Asx residue content of erythrocyte membrane proteins do not correspond to parallel differences in the mean age of circulating red blood cell populations between normal and uremic patients [62]. Mathematical modelling has indeed shown that this degree of inhibition of PCMT in uraemia may be responsible of an up to 600-fold increase of L-isoaspartyl residues in erythrocyte proteins, compared with normal erythrocytes, where the enzyme is instead able to counterbalance the formation of these damaged protein sites, thus preventing their accumulation [62].

It should be noted that the calculated racemized level (D-Asp + D-isoAsp) based on *S*-adenosylhomocysteine inhibition of PCMT matches the experimental value more closely than the agreement between the observed and calculated number of methylated residues (Figs 5 and 6). This suggests that the inhibition of PCMT-mediated repair of protein damage in uraemic patients is in fact more dramatic than that evaluated by the methylation assay performed in intact erythrocytes [62].

These data based on pathophysiological models confirm the view that efficient repair of the major form of altered aspartyl residues, i.e., L-isoaspartyl residues, is obtained *in vivo* at the expense of a much smaller increase of aspartyl racemization, mainly in the form of D-isoaspartyl residues.

Conclusions

Data so far obtained in the field of D-amino acid residues in erythrocyte proteins point to the following conclusions:

- D-amino acids in erythrocyte membrane proteins are mostly represented by D-aspartyl and D-isoaspartyl residues. These may spontaneously arise from the opening of a D-succinimidyl ring. The latter can be generated by two ways: (i) a completely spontaneous series of reactions whose first step involves either deamidation of susceptible L-asparaginyl residues or, less frequently, isomerization of L-aspartyl residues – both occur through an L-succinimidyl intermediate, which is in turn racemization-prone; (ii) a side mechanism of the repair pathway of L-isoaspartyl residues in proteins, mediated by PCMT, including the enzymatic methyl esterification of L-isoaspartyl residues followed by hydrolysis of esters through the same L-succinimide intermediate seen above.
- The human erythrocyte is a terminally differentiated cell which permanently loses the ability of synthesizing proteins *de novo*. Therefore, proteins tend to accumulate a degree of molecular fatigue damage during the course of the 120-day aging of cells in circulation. This damage includes the formation of racemized aspartyl sites, both from all-spontaneous reactions and from enzymatic methylation-mediated mechanisms. The functional consequences of the appearance of these altered residues in erythrocyte proteins, as well as in other tissues and

organs, is still under investigation. Nevertheles, data so far obtained indicate that D-Asx (and particularly D-isoaspartyl) accumulation is not just a mere marker of erythrocyte protein aging but also a reliable indicator of the activity of the protein methyltransferase repair enzyme, particularly with respect to certain crucial membrane protein components. In this respect the lessening of racemized aspartyl residues in proteins would represent a red flag in a disrupted protein repair circuit under conditions, such as uremia, in which the enzyme PCMT is significantly inhibited.

- The enzyme PCMT is not only able to stem the tide of increasing altered aspartyl residues, which are to some extent expected to occur in proteins during erythrocyte aging. Nevertheless, studies performed on model diseases, support the view that enzyme PCMT is like a highly specialized policeman on duty in surveilling and repairing specific protein damage resulting not only from erithrocyte aging, but also due to cell stress.

References

1. Wold, F. (1981) *In vivo* Chemical modifications of proteins. *Ann Rev Biochem* 50:783–814.
2. Harding, J.J. (1985) Nonenzymatic covalent posttranslational modifications of proteins *in vivo*. *Adv Protein Chem* 37:247–334.
3. Galletti, P., Ingrosso, D., Manna, C., Clemente, G. and Zappia, V. (1995) Protein damage and methylation mediated repair in the erythrocyte. *Biochem J* 306:313–325.
4. Capasso, S., Demasi, D. and Zagari, A. (1994) Deamidation of asparagine and isomerization of aspartic acid residues via succinimide derivatives: kinetic and structural studies on small peptides. *Curr Top Peptide Prot Res* 1:345–373.
5. Ota, I.M. and Clarke, S. (1990) The function and enzymology of protein D-aspartyl/L-isoaspartyl methyltrasferases in eukaryotic and prokaryotic cells. In: W.K. Paik and S. Kim (eds) *Protein Methylation*. CRC Press, Boca Raton, FL, 179–194.
6. Robinson, A.B. and Rudd, C.J. (1974) Deamidation of glutamyl and asparaginyl residues in peptides and proteins. *Curr Top Cell Regul* 8:247–295.
7. Bornsten, P. and Balian, G. (1977) Cleavage at Asn-Gly bonds with hydroxylamine *Methods Enzymol* 47:132–145.
8. Aswad, D.W. (1984) Stoichiometric methylation of procine adrenocorticotropin by protein carboxyl methyltransferase requires deamidation of asparagine 25. Evidence for methylation at the alpha-carboxyl group of atypical L-isoaspartyl residues. *J Biol Chem* 259: 10714–10721.
9. Murray, E.D. and Clarke, S. (1984) Synthetic peptide substrates for the erythrocyte protein carboxyl methyltransferase. Detection of a new site of methylation at isomerized L-aspartyl residues. *J Biol Chem* 259:10722–10732.
10. Di Donato, A., Galletti, P. and D'Alessio, G. (1986) Selective deamidation and enzymatic methylation of seminal ribonuclease. *Biochemistry* 25:8361–8368.
11. Capasso, S., Mazzarella, L., Sica, F. and Zagari, A. (1989) Deamidation *via* cyclic imide in asparaginyl peptides. *Peptide Res* 2:195–200.
12. Capasso, S., Mazzarella, L., Sica, F., Zagari, A. and Salvadori, S. (1993) Kinetics and mechanism of succinimide ring formation in the deamidation process of asparagine residues. *J Chem Soc Perkin Trans* 2:679–682.
13. Clarke, S. (1987) Propensity for spontaneous succinimide formation from aspartyl and asparaginyl residues in cellular proteins. *Int J Peptide Protein Res* 30:808–821.
14. Wright, T. (1991) Sequence and structure determinants for the nonenzymatic deamidation of asparagine and glutamine residues in proteins. *Protein Eng* 4:283–294.

15. Geiger, T. and Clarke, S. (1987) Deamidation, isomerization and racemization at asparaginyl and aspartyl residues in peptides. *J Biol Chem* 262:785–794.
16. Stephenson, R.C. and Clarke, C. (1989) Succinimide formation from aspartyl and asparaginyl peptides as a model for the spontaneous degradation of proteins. *J Biol Chem* 264: 6164–6170.
17. Ota, I.M. and Clarke, S. (1989) Calcium affects the spontaneous degradation of aspartyl/asparaginyl residues in calmodulin. *Biochemistry* 28:4020–4027.
18. Johnson, B.A., Langmack, E.L. and Aswad, D.W. (1987) Partial repair of deamidation-damaged calmodulin by protein carboxylmethyltransferase. *J Biol Chem* 262:12283–12287.
19. Ota, I.M. and Clarke, S. (1990) Multiple sites of methyl esterification of calmodulin in intact erythrocytes. *Arch Biochem Biophys* 279:320–327.
20. Desrosiers, R.R., Romanik, E.A. and O'Connor, C.M. (1990) Selective carboxyl methylation of structurally altered calmodulins in *Xenopus* oocytes. *J Biol Chem* 265:21368–21374.
21. Ota, I.M. and Clarke, S. (1989) Enzymatic methylation of L-isoaspartyl residues derived from aspartyl residues in affinity-purified calmodulin: the role of conformational flexibility in spontaneous isoaspartyl formation. *J Biol Chem* 264:54–60.
22. Galletti, P., Iardino, P., Ingrosso, D., Manna, C. and Zappia, V. (1989) Enzymatic methyl esterification of a deamidated form of mouse epidermal growth factor. *Int J Peptide Protein Res* 33:397–402.
23. Galletti, P., Ciardiello, A., Ingrosso, D., Di Donato, A. and D'Alessio, G. (1988) Repair of isopeptide bonds by protein carboxyl *O*-methyltransferase: seminal ribonuclease as a model system. *Biochemistry* 27:1752–1757.
24. Artigues, A., Birkett, A. and Schirch, V. (1990) Evidence for the *in vivo* deamidation and isomerization of an asparaginyl residue in cytosolic serine hydroxymethyltransferase. *J Biol Chem* 265:4853–4858.
25. Yuan, P.M., Talent, J.M. and Gracy, R.W. (1981) Molecular basis for the accumulation of acidic isozymes of triosephosphate isomerase on aging. *Mech Ageing Develop* 17: 151–161.
26. Friedman, A.R., Ichpurani, A.K., Brown, D.M., Hillman, R.M., Krabill, L.F., Martin, R.A., Zurcher-Neely, H.A. and Guido, D.M. (1991) Degradation of growth hormone-releasing factor analogs in neutral aqueous solution is related to deamidation of asparagine residues. Replacement of asparagine residues by serine stabilizes. *Int J Pept Protein Res* 37:14–20.
27. George-Nascimento, C., Lowenson, J.D., Borissenko, M., Calderon, M., Medina-Selby, A., Kuo, J., Clarke, S. and Randolph, A. (1990) Replacement of a labile aspartyl residue increases the stability of human epidermal growth factor. *Biochemistry* 29:9584–9591.
28. Yamada, H., Ueda, T., Juroki, R., Fukumura, T., Yasukochi, T., Hirabayashi, T., Fujita, K. and Imoto, T. (1985) Isolation and characterization of 101-beta-lysozyme that possesses the beta-aspartyl sequence at aspartic acid-101. *Biochemistry* 24:7953–7959.
29. Bonaventura, J., Bonaventura, C., Sullivan, B., Ferruzzi, G., McCurdy, P.R., Fox, J. and Moo-Penn, W.F. (1976) Hemoglobin providence. Functional consequences of two alterations of the 2,3-diphosphoglycerate binding site at position beta 82. *J Biol Chem* 251: 7563–7571.
30. Ota, I.M., Ding, L. and Clarke, S. (1987) Methylation at specific altered aspartyl and asparaginyl residues in glucagon by the erythrocyte protein carboxyl methyltransferase. *J Biol Chem* 262:8522–8531.
31. Johnson, B.A., Shirokawa, J.M., Hancock, W.S., Spellman, H.W., Basa, L.J. and Aswad, D.W. (1989) Formation of isoaspartate at two distinct sites during *in vitro* aging of human growth hormone. *J Biol Chem* 264:14262–14271.
32. Brennan, T.V. and Clarke, S. (1993) Spontaneous degradation of polypeptides at aspartyl and asparaginyl residues: effects of the solvent dielectric. *Protein Science* 2:331–338.
33. Capasso, S., Mazzarella, L. and Zagari, A. (1991) Deamidation *via* cyclic imide of asparaginyl peptides: dependence of salts, buffers and organic solvents. *Peptide Research* 4: 234–238.
34. Roher, A.E., Lowenson, J.D., Clarke, S., Wolkow, C., Wang, R., Cotter, R.J., Reardon, I.M., Zürcher-Neely, H.A., Heinrikson, R.L. and Ball, M.J. et al. (1993) Structural alterations in the peptide backbone of β-amyloid core protein may account for its deposition and stability in Alzheimer's disease. *J Biol Chem* 268:3072–3083.

35. Fabian, H., Szendrei, G.I., Mantsch, H.H., Greenberg, B.D. and Ötvös, L. (1994) Synthetic post-translationally modified human Aβ peptide exhibits a markedly increased tendency to form β-pleated sheets *in vitro*. *Eur J Biochem* 221:959–964.
36. Roher, A.E., Lowenson, J.D., Clarke, S., Woods, A.S., Cotter, R.J., Gowing, E. and Ball, M.J. (1993) β-amyloid-(1-42) is a major component of cerebrovascular amyloid deposits: implications for the pathology of Alzheimer disease. *Proc Natl Acad Sci USA* 90: 10836–10840.
37. Tomiyama, T., Asano, S., Furiya, Y., Shirasawa, T., Endo, N. and Mori, H. (1994) Racemization of Asp^{23} residue affects the aggregation properties of Alzheimer amyloid β protein analogues. *J Biol Chem* 269:10205–10208.
38. Hefman, P.M. and Bada, J.L. (1976) Aspartic acid racemization in dentine as a measure of aging. *Nature* 262:279–281.
39. Masters, P.M., Bada, J.L. and Zigler, J.S. (1978) Aspartic acid racemization in heavy molecular weight crystallins and water-insoluble protein from normal human lenses and cataracts. *Proc Natl Acad Sci USA* 75:1204–1208.
40. Lubec, G., Weniger, M. and Anderson, S.R. (1994) Racemization and oxidation studies of hair protein in the *Homo tirolensis*. *FASEB J* 8:1166–1169.
41. Brunauer, L.S. and Clarke, S. (1986) Age-dependent accumulation of protein residues which can be hydrolyzed to D-aspartic acid in human erythrocytes. *J Biol Chem* 261:12538–12543.
42. Lowenson, J.D. and Clarke, S. (1988) Does the chemical instability of aspartyl and asparaginyl residues in proteins contribute to erythrocyte aging? *Blood Cells* 14:103–117.
43. Galletti, P., Ingrosso, D., Manna, C., Sica, F., Capasso, S., Pucci, P. and Marino, G. (1988) Enzymatic methyl esterification of synthetic tripeptides: structural requirements of the peptide substrates. *Eur J Biochem* 177:233–239.
44. Lowenson, J.D. and Clarke, S. (1992) Recognition of D-aspartyl residues in polypeptides by the erythrocyte L-isoaspartyl/D-aspartyl protein methyltransferase. Implications for the repair hypothesis. *J Biol Chem* 267:5985–5995.
45. Lowenson, J.D. and Clarke, S. (1991) Structural elements affecting the recognition of L-isoaspartyl residues by the L-isoaspartyl/D-aspartyl protein methyltransferase. Implications for the repair hypothesis. *J Biol Chem* 266:19396–19406.
46. Ingrosso, D., Fowler, A.V., Bleibaum, J. and Clarke, S. (1989) Sequence of the D-aspartyl/L-isoaspartyl methyltransferase from human erythrocytes: common sequence motifs for protein, DNA, RNA and small molecule *S*-adenosylmethionine-dependent methyltransferases. *J Biol Chem* 264:20131–20139.
47. Ingrosso, D., Kagan, R.M. and Clarke, S. (1991) Distinct C-terminal sequences of isozymes I and II of the human erythrocyte L-isoaspartyl/D-aspartyl protein methyltransferase. *Biochem. Biophys Res Commun* 175:351–358.
48. Ingrosso, D. and Clarke, S. (1991) Human erythrocyte D-aspartyl/L-isoasparty methyltransferase: enzymes that recognize age-damaged proteins. In: M. Magnani and A. De Flora (eds): *Red Blood Cell Aging*. Plenum Press, New York, 263–276.
49. MacLaren, D.C., Kagan, R.M. and Clarke, S. (1992) Alternative splicing of the human isoaspartyl protein carboxyl methyltransferase RNA leads to the generation of a C-terminal -RDEL sequence in isoenzyme II. *Biochem Biophys Res Commun* 185:277–283.
50. MacLaren, D.C., O'Connor, C.M., Yia, Y.-R., Mehrabian, M., Klisak, V., Sparkes, R.S., Clarke, S. and Lusis, A.J. (1992) The L-isoaspartyl/D-aspartyl protein methyltransferase gene (PCMT1) maps to human chromosome 6q22.3–6q24 and the synthenic region of mouse chromosome 10. *Genomics* 14:1–5.
51. DeVry, C.G., Tsai, W. and Clarke, S. (1996) Structure of the human gene encoding the protein repair (L-isoaspartyl (D-aspartyl) *O*-methyltransferase. *Arch Biochem Biophys* 335:321–332.
52. David, C.L., Szumlanski, C.L., DeVry, C.G., Park-Hah, J.O., Clarke, S., Weinshilboum, R.M. and Aswad, D.W. (1997) Human erythrocyte protein L-isoaspartyl methyltransferase: heritability of basal activity and genetic polymorphis for thermal stability. *Arch Biochem Biophys* 346:277–286.
53. Kagan, R.N. and Clarke, S. (1994) Widespread occurrence of three sequence motifs in diverse *S*-adenosylmethionine-dependent methyltransferases suggests a common structure for these enzymes. *Arch Biochem Biophys* 310:417–427.
54. Paranandi, M.V., Guzzetta, A.W., Hancock, W.S. and Aswad, D.W. (1994) Deamidation and isoaspartate formation during in *in vitro* aging of recombinant tissue plasminogen activator. *J Biol Chem* 269:243–253.

55. Sharokh, Z., Eberlein, G., Buckley, D., Paranandi, M.V., Aswad, D.W., Stratton, P., Mischak R. and Wang, Y.J. (1994) Major degradation products of basic fibroblast growth factor: detection of succinimide and iso-aspartate in place of aspartate. *Pharm Res* 11:936–944.
56. Paranandi, M.V. and Aswad, D.W. (1995) Spontaneous alterations in the covalent structure of synapsin I during *in vitro* aging. *Biochem Biophys Res Commun* 212:442–448.
57. Najbauer, J., Orpiszewski, J. and Aswad, D.W. (1996) Molecular aging of tubulin: accumulation of isoaspartyl sites *in vitro* and *in vivo*. *Biochemistry* 35:5183–5190.
58. McFadden, P.N. and Clarke, S. (1987) Conversion of isoaspartyl peptides to normal peptides: implications for the cellular repair of damaged proteins. *Proc Natl Acad Sci USA* 84:2595–2599.
59. Johnson, B.A., Murray, E.D. Jr, Clarke, S., Glass, D.B. and Aswad, D.W. (1987) Protein carboxyl methyltransferase facilitates conversion of atypical L-isoaspartyl peptides to normal L-aspartyl peptides. *J Biol Chem* 262:5622–5629.
60. Brennan, T.V., Anderson, J.W., Jia, Z., Waygood, E.B. and Clarke, S. (1994) Repair of spontaneously deamidated HPr phosphocarrier protein catalyzed by the L-isoaspartate-(D-aspartate)-*O*-methyltransferase. *J Biol Chem* 40:24586–24595.
61. Lowenson, J.D. and Clarke, S. (1991) Spontaneous degradation and enzymatic repair of aspartyl and asparaginyl residues in aging red cell proteins analyzed by computer simulation. *Gerontology* 37:128–151.
62. Perna, A.F., D'Aniello, A., Lowenson, J.D., Clarke, S., De Santo, N.G. and Ingrosso, D. (1997) D-aspartate content of erythrocyte membrane proteins is decreased in uremia: implications for the repair of damaged proteins. *J Am Soc Nephrol* 8:95–104.
63. McFadden, P.N. and Clarke, S. (1986) Chemical conversion of aspartyl peptides to isoaspartyl peptides. A method for generating new methyl-accepting substrates for the erythrocyte D-aspartyl/L-isoaspartyl protein methyltransferase. *J Biol Chem* 261:11503–11511.
64. Fujii, N., Satoh, K., Harada, K. and Ishibashi, Y. (1994) Simultaneous stereoinversion and isomerization at specific aspartic acid residues in alpha A-crystallin from human lens. *J Biochem* 116:663–669.
65. Lindquist, J.A. and McFadden, P.N. (1994) Automethylation of protein (D-aspartyl/L-isoaspartyl carboxylmethyltransferase, a response to enzyme aging. *J Protein Chem* 13:23–30
66. Lindquist, J.A., Barofsky, E. and McFadden, P.N. (1996) Determination of two sites of automethylation in bovine erythrocyte protein (D-aspartyl/L-isoaspartyl) carboxyl methyltransferase. *J Protein Chem* 15:115–122.
67. Ladino, C.A. and O'Connor, C.M. (1992) Methylation of atypical protein aspartyl residues during stress response of HeLa cells. *J Cell Physiol* 153:297–304.
68. O'Connor, C.M. (1987) Regulation and subcellular distribution of a protein methyltransferase and its damaged aspartyl substrate sites in developing *Xenopous* oocytes. *J Biol Chem* 262:10398–10403.
69. Romanik, E.A. and O'Connor, C.M. (1989) Methylaton of microinjected isoaspartyl peptides in *Xenopous* oocytes. Competition with protein carboxyl methylation reaction. *J Biol Chem* 264:14050–14056.
70. Kagan, R.M. and Clarke, S. (1995) Protein L-isoaspartyl methyltransferase from the nematode *Caernorhabditis elegans*: genomic structure and substrate specificity. *Biochemistry* 34:10794–10806.
71. O'Connor, M.B., Galus, A., Hartenstine, M., Magee, M., Jackson, F.R. and O'Connor, C.M. (1997) Structural organization and developmental expression of the protein isoaspartyl methyltransferase gene from *Drosophila melanogaster*. *Insect Biochem Mol Biol* 27: 49–54.
72. Kim, E., Lowenson, J.D., MacLaren, D.C., Clarke, S. and Young, S.G. (1997) Deficiency of a protein-repair enzyme results in the accumulation of altered proteins, retardation of growth, and fatal seizures in mice. *Proc Natl Acad Sci USA* 94:6132–6137.
73. Weber, D.J. and McFadden, P.N. (1997) Detection and characterization of a protein isoaspartyl methyltransferase which becomes trapped in the extracellular space during blood vessel injury. *J Protein Chem* 16:257–267.
74. Weber, D.J. and McFadden, P.N. (1997) Injury-induced enzymatic methylation of aging collagen in the extracellular matrix of blood vessels. *J Protein Chem* 16:269–281.
75. Hassoun, H. and Palek, J. (1996) Hereditary spherocytosis: a review of the clinical and molecular aspects of the disease. *Blood Reviews* 10:129–147.

76. Magnani, M. and De Flora, A. (eds) (1991) *Red Blood Cell Aging*. Plenum Press, New York.
77. Haram, S., Carriero, D., Seaman, C. and Piomelli, S. (1991) The mechanism of decline of age-dependent enzymes in the red blood cell. *Enzyme* 45:47–53.
78. Barber, J.R. and Clarke, S. (1984) Inhibition of protein carboxyl methylation by *S*-adenosyl-L-homocysteine in intact erythrocytes. Physiological consequences. *J Biol Chem* 259: 7115–7122.
79. Perna, A.F., Ingrosso, D., Zappia, V., Galletti, P., Capasso, G. and De Santo, N.G. (1993) Enzymatic methyl esterification of erythrocyte membrane proteins is impaired in chronic renal failure. Evidence for high levels of the natural inhibitor *S*-adenosylhomocysteine. *J Clin Invest* 91:2497–2503.
80. Barber, J.R., Morimoto, B.H., Brunauer, L.S. and Clarke, S. (1986) Metabolism of *S*-adenosyl-L-methionine in intact human erythrocytes. *Biochim Biophys Acta* 886:361–372.
81. Freitag, C. and Clarke, S. (1981) Reversible methylation of cytoskeletal and membrane proteins in intact human erythrocytes. *J Biol Chem* 256:6102–6108.
82. O'Connor, C.M. and Clarke, S. (1984) Carboxyl methylation of cytosolic proteins in intact human erythrocytes. Identification of numerous methyl-accepting proteins including hemoglobin and carbonic anhydrase. *J Biol Chem* 259:2570–2578.
83. O'Connor, C.M. and Yutzey, K.E. (1988) Enhanced carboxyl methylation of membrane-associated hemoglobin in human erythrocytes. *J Biol Chem* 263:1386–1390.
84. Ladino, C.A. and O'Connor, C.M. (1991) Identification of a site for carboxyl methylation in human alpha-globin. *Biochem Biophys Res Common* 180:742–747.
85. Kimzey, A.L. and McFadden, P.N. (1994) Spontaneous methylation of hemoglobin by *S*-adenosyl-methionine by a specific and saturable mechanism. *J Protein Chem* 13:537–546.
86. Galletti, P., Ingrosso, D., Nappi, A., Gragnaniello, V., Iolascon, A. and Pinto, L. (1983) Increased methyl esterification of membrane proteins in aged red blood cells. Preferential esterification of ankyrin and band 4.1 cytoskeletal proteins. *Eur J Biochem* 135:25–31.
87. Barber, J.R. and Clarke, S. (1983) Membrane protein carboxyl methylation increases with human erythocyte age. Evidence for an increase in the number of methylatable sites. *J Biol Chem* 258:1189–1196.
88. Ladino, C.A. and O'Connor, C.M. (1990) Protein carboxyl methylation and methyl ester turnover in density-fractionated human erythrocytes. *Mech Ageing Dev* 55:123–137.
89. Inaba, M., Gupta, K.C., Kuwabara, M., Takahashi, T., Benz, E.J. Jr. and Maede, Y. (1992) Deamidation of human erythrocyte protein 4.1: Possible role in aging. *Blood* 79:3355–3361.
90. Ingrosso, D., D'Angelo, S., Perna, A.F., Miraglia del Giudice, E., Iolascon, A., Perrotta, S., Zappia, V. and Galletti, P. (1995) Increased membrane protein methylation in hereditary spherocytosis: a marker of cytoskeletal disarray. *Eur J Biochem* 228:894–898.
91. Ingrosso, D., D'Angelo, S., Perrotta, S., d'Urzo, G., Iolascon, A., Perna, A.F., Galletti, P., Zappia, V. and Miraglia del Giudice, E. (1996) Cytoskeletal behaviour in spectrin and in band 3 deficient spherocytic red cells: evidence for a differentiated splenic conditioning role. *Br J Haematol* 93:38–41.
92. Ingrosso, D., Cotticelli, M.G., D'Angelo, S., Buro, M.D., Zappia, V. and Galletti, P. (1997) Influence of osmotic stress on protein methylation in resealed erythrocytes. *Eur J Biochem* 244:918–922.
93. Perna, A.F., Ingrosso, D., Galletti, P., Zappia, V and De Santo, N.G. (1996) Membrane protein damage and methylation reactions in chronic renal failure. *Kidney Int* 50:358–366.
94. Perna, A.F., Ingrosso, D., De Santo, N.G., Galletti, P. and Zappia, V. (1995) Mechanism of erythrocyte accumulation of methylation inhibitor *S*-adenosylhomocystene in uremia. *Kidney Int* 47:247–253.
95. Perna, A.F., Ingrosso, D., De Santo, N.G., Galletti, P., Brunone, M. and Zappia, V. (1997) Metabolic consequences of folate-induced reduction of hyperhomocysteinemia in uremia. *J Am Soc Nephrol* 8:1899–1905.

D-Amino acids hydrolysing enzymes

D-Amino Acids in Sequences of Secreted Peptides of Multicellular Organisms
ed. by P. Jollès

D-Amino acid hydrolysing enzymes

R. Yamada and Y. Kera

Department of Environmental Systems Engineering, Nagaoka University of Technology, Nagaoka, Niigata 940-21, Japan

Summary. Only a few enzymes that hydrolyse peptide bonds involving D-amino acids effectively have been discovered and characterised in multicellular organisms. Mammalian renal dipeptidase hydrolyses various dipeptides with a D-amino acid only at the C-terminal with similar efficiency to their L-amino acid diastereomers, but not dipeptides with an N-terminal D-amino acid residue. Nor does the enzyme act on tripeptides. Dipeptides similar to those hydrolysed by the enzyme are also hydrolysed by cytosolic leucine aminopeptidase, but much less effectively than their L-amino acid diastereomers. Peptidyl-D-amino acid hydrolase from cephalopods has a somewhat broader substrate specificity than the renal dipeptidase and hydrolyses, as well, some dipeptides with a D-amino acid at the N-terminal. It also acts on larger peptides than dipeptides, albeit slowly. Carnosinase is specific to dipeptides containing L-His as the C-terminal residue, and hydrolyses D-Ala-L-His about as well as carnosine.

Introduction

In recent years, many biologically active peptides containing D-amino acids have been discovered in frog and several invertebrates as described in other chapters of this book, and their functions and mechanisms for racemization at specific amino acid residues have been targets of active research. In addition, aged proteins containing D-amino acids, also described in other chapters, have attracted increasing attention because of their possible implication in Alzheimer's disease. On the other hand, information about the *in vivo* degradation of these peptides is limited. This question has been investigated to some extent in aged proteins, but it is still not certain whether D-amino acid residues are finally liberated. Biologically active peptides appear to have been little studied from this vantage. This chapter briefly summarizes the results of studies related to the *in vivo* fate of D-amino acid-containing peptides and proteins, and describes enzymes which hydrolyse peptide bonds involving D-amino acids.

Studies on the fate of D-amino acids in peptides and proteins

In 1987, Man and Bada summarized accumulated evidence on lowered digestibility of food proteins containing D-amino acid residues, which were produced by racemization induced mainly by heat and alkali treatments during various kinds of food processing [1]. They suggested that proteins containing racemized amino acids would probably yield, after undergoing actions of digestive enzymes, low molecular weight peptides enriched in D-amino acids, and pointed out that extensive studies of D-amino acids in small peptides were required in order to ascertain their metabolic fate.

Reports in those days already provided evidence that peptides containing D-amino acids were resistant to hydrolysis by proteolytic enzymes. For example, Paquet et al. [2] examined the effect of the incorporation of D-amino acid residues into model tripeptides occurring in bovine β-casein on their hydrolysis by intestinal peptidases. They found that immobilized peptidases of porcine intestinal mucosa were inactive to L-Ala-D-Met-L-Ala and L-Val-D-Met-L-Phe, whereas their counterparts containing L-Met in place of D-Met were completely hydrolysed. Concerning the fate of aged proteins, Murray et al. synthesized a hexapeptide L-Val-L-Tyr-L-Pro-D-Asp-Gly-L-Ala, as a model compound, and treated the peptide with yeast carboxypeptidase Y or a mixture of porcine kidney cytosolic leucine aminopeptidase and prolidase [3]. No digestion was found with carboxypeptidase Y, since D-amino acid-containing peptides were competitive inhibitors of this enzyme. Digestion with the mixture of two enzymes resulted in the release of L-Val, L-Tyr, L-Pro and a fourth compound which was probably D-Asp-Gly-L-Ala, indicating that D-aspartyl residue was not liberated. Of the aged proteins, β-amyloid peptides containing D-amino acids have drawn much attention lately, since neuronal plaque amyloid from Alzheimer's disease brain tissues was found to contain significant levels of D-Asp and D-Ser [4]. Kaneko et al. studied the cytotoxic effects of various synthetic β-amyloids, including β25–35(25Gly-L-Ser-L-Asn-L-Lys-Gly-L-Ala-L-Ile-L-Ile-Gly-L-Leu-L-Met35) and its D-Ser-substituted derivative at position 26 [5]. Both synthetic β-amyloids exerted similar neurodegenerative effects on rat hippocampal cultured neurons as judged by cell counting, probably through suppression of mitochondrial succinate dehydrogenase. On the other hand, they differed clearly in their sensitivity to proteolytic activities. β25–35 was degraded rapidly by proteinases in rat and bovine brain extracts or leucine aminopeptidase, whereas D-Ser-substitued β25–35 was resistant. The authors suggest that racemization of serine residues may be involved in neurodegeneration and formation of senile plaques through escape from the degradation process by brain proteinases.

Supported by the recent development of combinatorial chemistry, a growing number of peptides composed partially or entirely of D-amino acids have been synthesized. The main purpose of synthesis seems to be to provide metabolically stable drug candidates. In fact, some of the peptides have proved to be potent, long-acting substitutes for naturally occurring peptide hormones and antibiotics [6–11]. Most of the peptides composed exclusively of D-amino acids have been described as metabolically stable, protease-resistant or undegradable. The work by Miller et al. represents extensive studies of this kind [12]. They prepared six peptides composed of four to six L-amino acids, which were designed to be good substrates of carboxypeptidase A, papain, pepsin, elastase, trypsin and chymotrypsin, respectively. They also prepared six homologous peptides composed of corresponding D-amino acids. While the L-amino acid peptides were readily cleaved by the enzymes, the D-amino acid peptides were cleaved minimally or not at all by the respective enzymes. *In vivo* study by Pappenheimer et al. showed that five different octapeptides entirely composed of D-amino acids were recovered almost quantitatively from urine and feces after either oral or parenteral administration to rats [13]. On the other hand, He et al. obtained evidence of gut metabolism of some oligopeptides comprising D-amino acids in their study of absorption of orally administered D-oligopeptides [14]. The researchers administered to rats orally and intravenously 11 oligopeptides containing [^{3}H]labelled D-Phe at the N-terminal and other D-amino acids at other positions, collected urine for 24 h and examined the total radioactivity of the urine and that associated with unchanged peptide. They found significant differences between the values based on the total and unchanged compound for some of the orally administered peptides. Of these peptides, D-Phe-D-Gln, D-Phe-D-Glu and D-Phe-D-Ala-D-Val-D-Ala were concluded to be partially metabolized in the gastrointestinal tract, since they had high radiolabelled purity, and there were no metabolites detected following intravenous administration. Moreover, in the case of D-Phe-D-Ala-D-Val-D-Ala, increasing the dose by a factor of 10,000 resulted in enhanced oral bioavailability, which was defined as the ratio of percentage in urine of unchanged peptide resulting from oral vs. intravenous administration. This was an additional suggestion that this peptide undergoes some metabolism, which might be saturable in the gastrointestinal tract. The observed metabolism was presumably caused by aminopeptidases, endopeptidases or dipeptidylaminopeptidases that reside on the apical surface of the intestinal epithelium. It should be added, however, that most of the other peptides tested, including D-Phe-D-Ala, were absorbed without metabolism, and all the peptides were relatively stable following intravenous administration, with generally greater than 60% excreted unchanged.

Some peptides that partially contain D-amino acids are partially degradable, but reports of liberation of D-amino acids are rare. Kertscher et al. studied buserlin degradation pathways via rat kidney membrane [15]. Buserelin [L-pGlu-L-His-L-Trp-L-Ser-L-Tyr-D-Ser(tBu)-L-Leu-L-Arg-L-Pro-ethylamide] is one of the most frequently used superactive agonists of hypothalamic decapeptide hormone (L-pGlu-L-His-L-Trp-L-Ser-L-Tyr-Gly-L-Leu-L-Arg-L-Pro-Gly-NH_2). They observed that L-Tyr-D-Ser(tBu)-L-Leu-L-Arg-L-Pro-ethylamide appeared rapidly as one of the degradation products, and that this peptide was slowly converted to D-Ser(tBu)-L-Leu-L-Arg-L-Pro-ethylamide with the liberation of L-Tyr, showing that the presence of a D-amino acid in the N-terminal peptide bond did not prevent cleavage, but no further degradation was observed during 24-h incubation.

Similarly, Su et al. studied the degradative process of the anti-acquired immunodeficiency syndrome (AIDS) drug, [D-Ala^1]Peptide T amide(D-

Ala-L-Ser-L-Thr-L-Thr-L-Thr-L-Asn-L-Tyr-L-Thr-NH_2) in the small intestine [16]. They examined the metabolism of the peptide in rabbit brush-border membrane vesicles; the shortest peptide containing D-Ala they found under the experimental conditions was D-Ala-L-Ser-L-Thr-L-Thr. The work by Lister et al. represents one of the rare reports demonstrating that D-amino acids can be liberated from dipeptides in animal tissues [17]. The researchers studied transport and hydrolysis of dipeptides in isolated loops of rat small intestine. The dipeptides examined were eight dipeptides formed from D- and L-Ala plus D- and L-Phe. Isolated jejunal loops were perfused by a single pass of bicarbonate Krebs-Ringer solution containing one of the peptides at 0.1 – 1 mM concentrations, and serosal appearance of Phe and peptides was followed. When the peptide tested was L-Phe-L-Ala or L-Ala-L-Phe, Phe, whose configuration could not be determined by the method employed, appeared rapidly in the serosal secretions while no peptide appeared, indicating that the two peptides were completely hydrolysed. On the other hand, perfusion with D-Phe-D-Ala and D-Ala-D-Phe resulted in no appreciable serosal appearance of Phe and peptides, showing that these peptides were poor substrates for hydrolysis and transport. It should be noted that the metabolic stability of D-Phe-D-Ala has also been shown by *in vivo* experiments as described above. On perfusion with L-Ala-D-Phe, serosal Phe appearance was as rapid as in the case of L-Ala-L-Phe, and no appearance of the peptide was observed. L-Phe-D-Ala gave a similar pattern, with serosal appearance of Phe and not of the peptide, although the appearance was slower than in the case of L-Ala-D-Phe. When D-Ala-L-Phe and D-Phe-L-Ala were tested, both Phe and the peptides appeared in the serosal secretions, indicating that the peptides were partially hydrolysed. Data showed that D-Ala-L-Phe was a better substrate for both transport and hydrolysis. Thus, dipeptides containing one D-amino acid were hydrolysable by the enzymes associated with rat small intestine, and of these dipeptides, those having a C-terminal D-amino acid appeared to be better substrates than those with an N-terminal D-amino acid.

Enzymes that hydrolyse peptide bonds involving D-amino acids

The literature on these types of enzymes is still meager. Information available is limited to the following four enzymes.

Cytosolic leucine aminopeptidase (EC 3.4.11.1)

To our knowledge, cytosolic leucine aminopeptidase purified from porcine kidney is the first enzyme reported to be reactive to D-amino acid-containing peptides [18]. The enzyme has broad substrate specificity and hydrolyses not only leucyl compounds, as is suggeted by its name, but a wide variety of other amino acid amides and peptides. The enzyme was found to hydrolyse L-Leu-D-Leu, L-Leu-D-Ile, L-Leu-D-Val, L-Leu-D-Ala and L-Leu-D-Phe at a modest rate, with L-Leu-D-Ala being the best substrate. These peptides were, however, poor substrates in comparison with their counterparts containing a corresponding L-amino acid in place of the D enantiomer, and their reaction rate was, at best, a few percent of that of corresponding diastereomers. Later, it was further shown that Gly-D-Leu, Gly-D-Ala, Gly-D-Asp and Gly-D-Ser were also hydrolysed at a modest rate [19]. The enzyme did not hydrolyse dipeptides with a D-amino acid at the N-terminal position, suggesting that it was strict to the configuration of the N-terminal amino acid.

Carnosinase (aminoacyl-histidine dipeptidase EC 3.4.13.3)

Carnosinase partially purified from porcine kidney has been shown to hydrolyse D-Ala-L-His comparably well as the physiological substrate, carnosine (β-alanyl-L-histidine) [20]. The enzyme is specific to dipeptides containing L-His as the C-terminal residue.

Renal membrane dipeptidase (EC 3.4.13.19)

Renal membrane dipeptidase was first purified to homogeneity in 1965 from porcine kidney particulate material and was shown to hydrolyse a variety of dipeptides including Gly-D-Leu [21]. Later studies have shown that the enzyme acts on other peptides containing D-amino acids, that is Gly-D-Ala [22] and Gly-D-Phe [23]. A recent report by Watanabe et al. [24] has shown that the enzyme hydrolyses many more different dipetides containing D-amino acids, some of which are better substrates than their diastereomers containing L-amino acids. The researchers first found a hydrolytic activity toward Gly-D-Asp, which might appear as an intermediate in degradative pathway of aged proteins containing D-aspartyl residues, in porcine kidney cortex [25]. Purification of the enzyme responsible for the hydrolysis from brush-border membranes revealed that the enzyme, which had a molecular weight of 99,000, was a dimer of a subunit with a molecular weight of 48,000; the activity was highest at pH 7.8 and potently inhibited by cilastatin, but not by amastatin or bestatin; metal ion chelators and dithiothreitol were also inhibitory. These properties were almost the same as those reported for porcine renal membrane dipeptidase [21, 23, 26–29], with which the enzyme was therefore tentatively identified. The identification was further supported by later work (unpublished observations) showing that the sequence of the 30 N-terminal amino acid residues was the same as previously reported for renal membrane dipeptidase [30].

The enzyme was found to be active toward many dipeptides given in Table 1 but did not hydrolyse most of the tripeptides examined, which included D-Ala-Gly-Gly, L-Ala-Gly-Gly, L-Ala-L-Pro-Gly, D-Ala-L-Pro-L-Phe, L-Ala-L-Ala-L-Ala, D-Ala-D-Ala- D-Ala and D-Ala-L-Ala-L-Ala, as expected for a dipeptidase. Exeptionally, D-Ala-Gly-Gly and L-Ala-Gly-Gly were significantly hydrolysed at about 6% of the rate of Gly-D-Ala hydrolysis. The amino acid liberated from these peptides was Gly and not D- or L-Ala, suggesting that the enzyme acts as a carboxypeptidase.

Of the dipeptides examined, those having a D-amino acid residue at the N-terminal were poor substrates. Some, such as D-Ala-Gly, D-Leu-Gly and D-Leu-L-Tyr were not hydrolysed at all, and others, such as D-Leu-L-Leu, D-Leu-D-Leu and D-Ala-L-Leu, were markedly poorer substrates than their counterparts having a corresponding L-amino acid residue at the N-terminal. D-Ala-D-Ala could not be compared with L-Ala-D-Ala, but it was a poor substrate compared with L-Ala-L-Ala. In addition, kinetic analysis showed that K_m values for D-Leu-L-Leu and D-Leu-D-Leu were much higher than those for their counterparts. The findings probably mean that the enzyme does not bind well the depeptides bearing a D-amino acid residue at the N-terminal. An interesting exception was γ-D-Glu-Gly, which was hydrolysed more rapidly than γ-L-Glu-Gly. Perhaps the configuration around the carbon atom bonded to the carbonyl group of peptide bond is critical for binding. It should be pointed out that γ-D-Glu-*p*-nitroanilide has been shown to be hydrolysed by γ-glutamyl transpeptidase [31].

In contrast, the dipeptides having a D-amino acid residue situated at the C-terminal were hydrolysed well in comparison with their counterparts having an L-amino acid residue at the same position. The K_m value for Gly-D-Ala was remarkably lower (1/30) than that for Gly-L-Ala. Some of the dipeptides examined did not simply obey Michaelis-Menten kinetics. The Lineweaver-Burk plot of the data for Gly-L-Leu hydrolysis apparently contained two linear portions, and thus, kinetic parameters were tentatively calculated with these linear portions and are given in Table 1 for comparison. A similar phenomenon was observed for Gly-L-Phe, and the data were treated in the same way. These unusual plots might be due to substrate activation or other related mechanisms. In contrast to these peptides, Gly-D-Leu and Gly-D-Phe did not show this complicated behavior. Comparison of the K_m values listed in Table 1 suggests that dipeptides with a C-terminal D-amino acid bind to the enzyme as well as the corresponding L-amino acid peptides. However, V_{max} values were generally lower (1/2–1/8) for the D-amino acid peptides than for the corresponding L-amino acid peptides, with the exception of those for Gly-D-Ser and Gly-L-Ser, which were almost the same. The resultant V_{max}/K_m value was higher for the D-amino acid peptide than its L counterpart in some pairs, but lower in other pairs, and almost equal in still others, showing that there was no definite preference of one of the diastereomer series. The comparable activity thus demonstrated toward the diasteromeric dipeptides with either a D- or

Table 1. Activity of porcine renal dipeptidase toward various dipeptides

Substrate	Relative activity[a] (at 30 mM)	K_m (mM)	V_{max} units/mg protein	$\frac{V_{max}}{K_m}$	Turnover number[b] s^{-1}
Gly-D-Ala	100	2.7	660	240	520
Gly-L-Ala	74.1[c]	77	1600	21	1300
D-Ala-Gly		(not hydrolyzed)			
L-Ala-Gly	76.3[c]	0.51	470	930	380
Gly-D-Asp	8.0[c]	17	75	4.5	60
Gly-L-Asp	18.3	14	160	11	130
L-Asp-Gly	1.6	0.11	10	89	8.1
Gly-D-Ser	81.3	75	1600	21	1300
Gly-L-Ser	85.3	80	1800	23	1400
Gly-D-Phe	26.4[c]	3.8	180	47	140
Gly-L-Phe	122.7				
(1–15 mM)		4.9	670	140	530
(25–60 mM)		35	1600	46	1300
Gly-D-Leu	47.3[c]	3.5	320	92	250
Gly-L-Leu	141.9				
(0.4–20 mM)		3.4	620	180	500
(30–90 mM)		61	2000	33	1600
D-Leu-Gly		(not hydrolyzed)			
L-Leu-Gly	81.1[c]	0.53	500	950	400
L-Leu-D-Leu	3.5[c]	0.15	21	140	17
L-Leu-L-Leu	27.9[c]	0.16	170	1100	140
D-Leu-L-Leu	2.6[c]	640	350	0.55	280
D-Leu-D-Leu	19.0[c]	77	410	5.3	330
L-Ala-L-Ala	182.4				
D-Ala-D-Ala	2.7				
L-Ala-L-Leu	70.5				
D-Ala-L-Leu	0.9				
L-Leu-L-Tyr	13.5				
D-Leu-L-Tyr	0.0				
Gly-L-Tyr	129.1				
Gly-L-Asn	101.2				
Gly-D-Asn	31.9				
Gly-L-Gln	104.9				
Gly-L-Glu	9.5				
Gly-L-Thr	36.2				
Gly-D-Thr	8.7				
Gly-L-Val	9.2				
Gly-D-Val	0.0				
Gly-L-Ile	4.9				
Gly-L-Met	194.9				
Gly-L-His	74.9				
Gly-L-Lys	57.6				
Gly-L-Pro	4.6				
γ-D-Glu-Gly	19.6				
γ-L-Glu-Gly	5.3				
γ-L-Gly-L-Glu	7.0				
L-Gly-L-Glu	110.2				
β-L-Asp-Gly	2.7				

The results are the means for two experiments. The amino acids underlined were determined by HPLC. [a] The activities were expressed relative to Gly-D-Ala. [b] Turnover number was calculated based on the molecular mass of the monomer subunit (48 kDa). [c] The relative activity was calculated based on Michaelis-Menten equations. Reprinted from ref. 24 with the kind permission of Elsevier Science, Amsterdam, The Netherlands.

L-amino acid at the C-terminal suggests the possibility that the enzyme in fact plays a role in the *in vivo* degradation of D-amino acid-containing peptides.

Consistent with this suggestion the enzyme has been shown to be widely distributed in various organs of rat and other mammals: Hirota et al. [32] reported that the activity of dehydropeptidase-I, which is generally accepted today to be identical with the renal membrane dipeptidase, was highest in rat lung followed by kidney, spleen and liver, brain and pancreas, and small intestine, complementary DNA (cDNA) of this enzyme has been cloned from various mammals including human, porcine, mouse and rabbit [33 and references therein].

Peptidyl-D-amino acid hydrolase (EC 3.4.13.17) from *Loligo vulgaris* Lam.

While the enzymes described above are all from mammalian tissues, the enzyme decribed here is from cephalopod tissue and has a broader specificity to peptides that have a low number of D- or L-amino acids. Based on the presence of free D-aspartic acid in some organs of cephalopods [34, 35], D'Aniello et al. postulated that cephalopods would possess a hydrolase able to cut the peptide bonds of D-amino acid peptides. In fact, they found the enzyme, which they tentatively called peptidyl-D-amino acid hydrolase in various organs of the animals, and purified the enzyme to homogeneity from the richest source, digestive juice from cecum intestine of *Loligo vulgaris* [18]. The enzyme has a molecular weight of 140,000 with an isoelectric point at pH 6.1, and consists of two subunits with molecular weights of 106,000 and 36,000; pH optimum is at 8.0; EDTA and ordinary metal ions do not affect its activity.

Of the peptides containing D-amino acids examined, L-Ala-D-Ala was the best substrate and was completely hydrolysed, as was L-Ala-L-Ala under certain conditions (30 min of incubation at 37°C with 3.8 μg of enzyme). Dipeptides with a D-amino acid situated at the C-terminal, such as Gly-D-Leu, Gly-D-Ala, Gly-D-Asp, Gly-D-Phe, Gly-D-Ser and L-Tyr-D-Arg, were good substrates, comparable to their counterparts with an L-amino acid at the same position, and were hydrolysed 10–90% under the same conditions. An exception was L-Leu-D-Leu, which underwent only 2% hydrolysis. Of the dipeptides bearing a D-amino acid at the N-terminal, D-Ala-L-Ala, D-Ala-D-Ala and D-Leu-Gly underwent only a few percent hydrolysis, whereas D-Leu-L-Leu, D-Leu-D-Leu, D-Leu-L-Tyr and D-Ala-Gly were hydrolysed 15–75%, demonstrating the broader specificity of this enzyme than that of the renal membrane peptidase described above. The enzyme hydrolysed peptides composed of three or more amino acids slowly whether or not they contained D-amino acids, acting as a carboxypeptidase. It is notable that the enzyme liberated D-Ala from [D-Ala^2]-Met-enkephalin (L-Tyr-D-Ala-Gly-L-Phe-L-Met). The role of the enzyme seems likely to digest ingested foods in view of the finding that the digestive juice was the richest source of the enzyme and the recent discovery of peptides containing D-amino acids in various animals which could serve as food sources for cephalopods.

Discussion and future perspectives

Information presented on the biological degradation of peptides and proteins containing D-amino acids is mostly limited to mammals with the exception of cephalopods, while interesting biologically active peptides containing D-amino acids have been recently discovered.

In mammalian bodies, peptides entirely composed of D-amino acids mostly appear to be metabolically stable. These types of peptides consisting of three or more residues have been shown to be intact, although a tetrapeptide of this type showed evidence of limited gut metabolism in the rat. Some of these dipeptides, e.g., D-Leu-D-Leu, are hydrolysed by the renal membrane dipeptidase and some provide evidence of partial metabolism in gastrointestinal tract, whereas others, including D-Phe-D-Ala, appear to be intact. In any event, these kinds of peptides are unlikely to be endogenous.

Peptides composed partially of D-amino acids have often been shown to be degradable at least to some extent. However, it is still unclear whether the D-amino acid residue is finally liberated or not, especially for peptides consisting of three or more residues. *In vivo* studies comparing the excretion of these peptides and their metabolites with that of intact peptides composed entirely of D-amino acids will provide usuful information. Of dipeptides containing one D-amino acid, those with the C-terminal D enantiomer are substrates of renal membrane dipeptidase and leucine amino peptidase, and therefore likely to be hydrolysed *in vivo*. Studies with isolated rat intestinal loops on absorption and hydrolysis of L-Ala-D-Phe and L-Phe-D-Ala have provided supporting evidence. In similar experiments, D-Ala-L-Phe and D-Phe-L-Ala were shown to be partially hydrolysed, whereas the enzymes described above do not act well on dipeptides with a D-amino acid at the N-terminal. The enzyme responsible for the hydrolysis will be identified by examining intestinal peptidases.

The substrate specificity of peptidyl-D-amino acid hydrolase from cephalopods is a little broader than that of renal membrane dipeptidase and hydrolyses dipetides with a D-amino acid at the N-terminal fairly well. It acts on tripeptides, although slowly, and liberates D-Ala from L-Tyr-D-Ala-Gly-L-Phe-L-Met. Thus it is very likely that the enzyme plays a significant role *in vivo*. The discovery of the enzyme suggests the possibility that similar enzymes will be found in frogs and various invertebrates which produce biologically active peptides containing D-amino acids. Degradation pathways of these peptides and the enzymes involved are expected to be rapidly elucidated with the well-developed materials and techniques currently available.

References

1. Man, E.H. and Bada, J.L. (1987) Dietary D-amino acids. *Ann Rev Nutr* 7:209–225.
2. Paquet, A., Thresher, W.C., Swaisgood, H.E. and Catignani, G.L. (1985) Syntheses and digestibility determination of some epimeric tripeptides occurring in dietary proteins. *Nutr Res* 5:891–901.
3. Murray, E.D. Jr. and Clarke, S. (1984) Synthetic peptide substrates for the erythrocyte protein carboxyl methyltransferase. *J Biol Chem* 259:10722–10732.
4. Shapira, R., Austin, G.E. and Mirra, S.S. (1988) Neuritic plaque amyloid in Alzheimer's disease is highly racemized. *J Neurochem* 50:69–74.
5. Kaneko, I., Yamada, N., Sakuraba, Y., Kamenosono, M. and Tutumi, S. (1995) Suppression of mitochondrial succinate dehydrogenase, a primary target of β-amyloid, and its derivative racemized at Ser residue. *J Neurochem* 65:2585–2593.
6. Bauer, W., Briner, U., Doepfner, W., Haller, R., Huguenin, R., Marbach, P., Petcher, T.J. and Pless, J. (1982) A very potent and selective octapeptide analogue of somastatin with prolonged action. *Life Sci* 31:1133–1140.
7. Lundin, S. and Vilhardt, H. (1986) Absorption of intragastrically administered DDAVP in conscious dogs. *Life Sci* 38:703–709.
8. Fuessl, H.S., Domin, J., Bloom, S.R. (1987) Oral absorption of somatostatin analogue SMS201-995: theoretical and practical implications. *Clin Sci* 72:255–257 .
9. Dooley, C.T., Chung, N.N., Wilkes, B.C., Schiller, P.W., Bidlack, J.M., Pasternak, G.W. and Houghten, R.A. (1994) An all D-amino acid opioid peptide with central analgesic activity from a combinatorial library. *Science* 266:2019–2021.
10. Merrifield, R.B., Juvvadi, P., Andreu, D., Ubach, J., Boman, A. and Boman, H.G. (1995) Retro and retroenantio analogs of cecropin-melittin hybrids. *Proc Natl Acad Sci USA* 92: 3449–3453.
11. Quillan, J.M., Jayawickreme, C.K. and Lerner, M.R. (1995) Combinatorial diffusion assay used to identify topically active melanocyte-stimulating hormone receptor antagonists. *Proc Natl Acad Sci USA* 92:2894–2898.
12. Miller, S.M., Simon, R.J., Ng, S., Zuckermann, R.N., Kerr, J.M. and Moos, W.H. (1995) Comparison of the proteolytic susceptibilities of homologous L-amino acid, D-amino acid, and N-substituted glycine peptide and peptoid oligomers. *Drug Dev Res* 35:20–32.
13. Pappenheimer, J.R., Karnovsky, M.L. and Maggio, J.E. (1997) Absorption and excretion of undegradable peptides: role of lipid solubility and net charge. *J Pharmcol Exp Ther* 280: 292–300.
14. He, Y.-L., Murby, S., Gifford, L., Collett, A., Warhurst, G., Douglas, K.T., Rowland, M. and Ayrton, J. (1996) Oral absorption of D-oligopeptides in rats via the paracellular route. *Pharm Res* 13:1673–1678.
15. Kertscher, U., Brudel, M., Mehlis, B., Sandow, J. and Beerger, H. (1995) Pathways of degradation of buserelin by rat kidney membrane. *J Pharmacol Exp Ther* 273:709–715.
16. Su, S.-F. and Amidon, G.L. (1995) Investigation into the intestinal metabolism of [D-Ala1^1] peptide T amide: implication for oral drug delivery. *Biochim Biophys Acta* 1245:62–68.
17. Lister, N., Sykes, A.P., Bailey, P.D., Boyd, C.A.R. and Bronk, J.R. (1995) Dipeptide transport and hydrolysis in isolated loops of rat small intestine: effects of stereospecificity. *J Physiol* 484:173–182.
18. Smith, E.L. and Spackman, D.H. (1955) Leucine aminopeptidase V. Activation, specificity, and mechanism of action. *J Biol Chem* 212:271–299.
19. D'Aniello, A. and Strazzullo, L. (1984) Peptidyl-D-amino acid hydrolase from *Loligo vulgaris* Lam. Purification and characterization. *J Biol Chem* 259:4237–4243.
20. Smith, E.L. (1955) Dipeptidases A. Carnosinase. *Methods Enzymol* 2:93–96.
21. Campbell, B.J., Lin, Y.-C., Davis, R.V. and Ballew, E. (1966) The purification and properties of a particulate renal peptidase. *Biochim Biophys Acta* 118:371–386.

22. Kozak, E.M. and Tate, S.S. (1982) Glutathione-degrading enzymes of microvillus membranes. *J Biol Chem* 257:6322–6327.
23. Hooper, N.M., Low, M.G. and Turner, A.J. (1987) Renal dipeptidase is one of the membrane proteins released by phosphatidylinositol-specific phopholipase C. *Biochem J* 244:465–469.
24. Watababe, T., Kera, Y., Matsumoto, T. and Yamada, R. (1996) Purification and kinetic properties of a D-amino-acid peptide hydrolysing enzyme from pig kidney cortex and its tentative identification with renal membrane dipeptidase. *Biochim Biophys Acta* 1298: 109–118.
25. Kera, Y., Funabashi, K., Matsumoto, T., Watanabe, T., Nagasaki, H. and Yamada, R. (1996) D-Aspartyl residue in a peptide can be liberated and metabolized by pig kidney enzymes. *Amino Acids* 10:187–196.
26. Armstrong, D.J., Mukhopadhyay, S.K. and Campbell, B.J. (1974) Physicochemical characterization of renal dipeptidase. *Biochemistry* 13:1745–1750.
27. René, A.M. and Campbell, B.J. (1969) Amino acid compostion and effect of pH on the kinetic parameters of renal dipeptidase. *J Biol Chem* 244:1445–1450.
28. Kim, H.S. and Campbell, B.J. (1983) Association of renal dipeptidase with the triton-insoluble fraction of kidney microvilli. *J Membrane Biol* 75:115–122.
29. Littlewood, G.M., Hooper, N.M. and Turner, A.J. (1989) Ectoenzyme of the kidney microbillar membrane. Affinity purification, chracterization and localization of the phospholipase C-solubilized form of renal dipeptidase. *Biochem J* 257:361–367.
30. Rached, E., Hooper, N.M., James, P., Semenza, G. and Turner, A.J. (1990) cDNA cloning and expression in *Xenopus laevis* oocytes of pig renal dipeptidase, a glycosyl-phosphatidylinositol-anchored ectoenzyme. *Biochem J* 271:755–760.
31. Tate, S.S. and Meister, A. (1985) γ-Glutamyl transpeptidase from kidney. *Methods Enzymol* 113:400–419.
32. Hirota, T., Nishikawa, Y., Tanaka, M., Igarashi, T. and Kitagawa, H. (1986) Characterization of dehydropeptidase-I in the rat lung. *Eur J Biochem* 160:521–525.
33. Satoh, S., Keida, Y., Konta, Y., Maeda, M., Matsumoto, Y., Niwa, M. and Kohsaka, M. (1993) Purification and molecular cloning of mouse renal dipeptidase. *Biochim Biophys Acta* 1163:234–242.
34. D'Aniello, A. and Giuditta, A. (1977) Identification of D-aspartic acid in the brain of Octopus vulgaris Lam. *J Neurochem* 29:1053–1057.
35. D'Aniello, A. and Giuditta, A. (1978) Presence of D-aspartate in squid axoplasm and in other regions of the cephalopod nervous system. *J Neurochem* 31:1107–1108.

Homochirality and life

D-Amino Acids in Sequences of Secreted Peptides of Multicellular Organisms
ed. by P. Jollès

Homochirality and life

W. A. Bonner

Department of Chemistry, Stanford University, Stanford, CA 94305, USA

Summary. After clarifying the freqently misused term *homochirality*, the crucial importance of homochirality and chiral purity in the development and maintenance of the essential biopolymers of life – proteins and nucleic acids – is discussed. The harsh and forbidding prebiotic environment during the era of cometary impact after formation of the Earth ~4.5 Gyr ago is described, after which the most important abiotic mechanisms proposed historically for the genesis of chiral molecules on the primitive Earth are enumerated. *Random* and *determinate* terrestrial mechanisms are each evaluated with regard to the environmental restraints imposed during the impact era, and it is concluded that all such mechanisms would be inapplicable and implausible in the realistic prebiotic environment. To circumvent these limitations, an extended hypothesis is presented describing an extraterrestrial source of homochiral terrestrial molecules. Illustrated in Figure 2, this scenario involves the partial asymmetric photolysis of the racemic constituents of organic mantles on interstellar dust grains by the circularly polarized ultraviolet components of the synchrotron radiation emanating from the neutron star remnants of supernovae. The resulting homochiral constituents with low enanantiomeric excesses (e.e.s) so produced in the organic mantles are subsequently conveyed to Earth either by direct accumulation or, more likely, after coalescence into comets or asteroids, followed by repetitive impingement during the impact era. Finally, the low e.e.s of the extraterrestrial homochiral molecules so introduced are amplified by terrestrial autocatalytic or polymerization mechanisms into a state of chiral purity, then are ultimately concentrated and protected by sequestration in the interiors of spontaneously formed protocellular vesicles – there to await further chemical evolution toward the biomolecules of life. Recent observations of the excess of L-over D-amino acids in the Murchison meteorite are cited as validation for the early stages of the proposed hypothesis.

Introduction

Chirality, optical activity and homochirality: Some definitions

Chiral objects are three-dimensional structures associated with an intrinsic right- or left-handedness, or *chirality*. This is seen, for example, in right- and left-handed screws, spiral staircases and organic molecules containing one or more asymmetric tetrahedral carbon atoms (C*). Molecular chirality may be illustrated by the two nonsuperposable mirror-image structures (*enantiomers*) of alanine, Ala, whose D and L enantiomers are shown in Figure 1. Enantiomers are *optically active*, rotating the plane of polarization of plane-polarized light. One enantiomeric *optical isomer* causes a (+) *dextro*-rotation and the other an equal and opposite (–) *levo*-rotation, as seen for the specific rotations, $[\alpha]^{25}_{D}$, of the Ala enantiomers in Figure 1. Optical isomers with more than one C* and which are not mirror images are called *diastereomers*. A 50:50 *racemic mixture* (*racemate*) of two enantiomers, e.g., DL-Ala, is optically inactive. Mixtures of enantio-

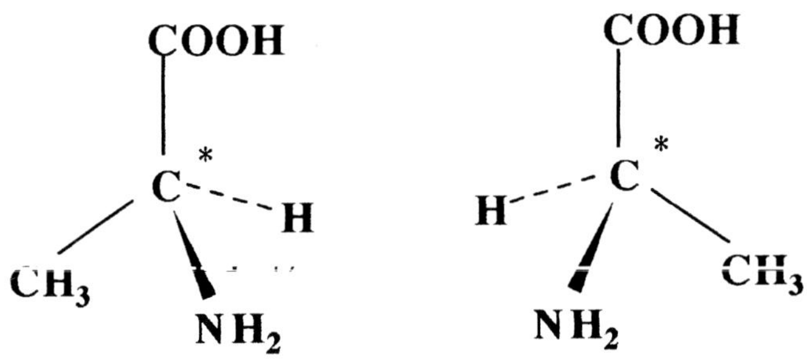

L(+)-Ala D(-)-Ala

$[a]^{25}_{D} = +14.5°$ $[a]^{25}_{D} = -14.5°$

Figure 1. Enantiomers of alanine.

mers show specific rotations which are intermediate between $[\alpha]^{25}_{D} = 0.0°$ for the racemate and that of the optically pure enantiomer. Such mixtures are conveniently characterized by their *enantiomeric excess* (e.e.), where e.e. = (% major enantiomer – % minor enantiomer) in the mixture. Thus the e.e. for an optically (i.e., enantiomerically) pure compound is 100%, whereas that of a racemate is 0%. The separtion of a racemate into its two enantiomers is referred to as *resolution*: the conversion of one enantiomer into a racemate is known as *racemization*. A *prochiral* molecule is one which may be converted into a chiral one by a suitable chemical reaction.

The frequently misunderstood and misused term *homochirality* applies to the chiral similarity of a group of externally similar "real objects" capable of enantiomerism [1]. In a molecular context the term implies the identical absolute stereochemical configurations of an array of chiral molecules which are structurally identical or similar. Thus the n monomer units of an $(\text{L-Ala})_n$ polypeptide are homochiral, as are the configurationally similar L-amino acids in a natural protein molecule. Note that it is the *array* of monomer units in these polymers, not the polymer molecules themselves, to which the therm *homochiral* correctly applies. When one speaks loosely of homochiral proteins, DNAs or RNAs, it is tacitly understood that it is the monomer units of these polymers which are homochiral and not the individual polymer molecules. We emphasize that these concepts have nothing whatsoever to do with optical activity, since the identical handedness of a group of similar molecules, not their optical activity, is the criterion for homochirality. Note also that "homochiral" should never be used to designate "optically pure" or "enantiomerically pure" (100% e.e.). These distintions are important because of current erroneous claims [2] that "Optical activity… offers on entirely general signature of homochirality" and that "extraterrestrial homochirality" may be recognized by the mere detection of small optimal rotations measured by rudimentary polarimeters carried aboard unmanned spacecraft. In the mid-1960s Halpern et al. had already suggested (the "Pasteur probe") that optical activity would provide a potential indicator for extraterrestrial life, but emphasized at the same time the inapplicability of polarimetric measurements for the purpose [3]. We shall see below that optical activity (i.e., an e.e.) *per se* is merely a necessary, but not a sufficient, criterion for life elsewhere in the cosmos.

In the following sections we have generally provided only the most recent pertinent references. For earlier and more extensive treatments of the topics considered, the reader is referred to our previous comprehensive reviews [4–7]. Page references in these reviews for the topics discussed below are indicated, for example, as ([4](18–20); [5](70–73); [6](40–43)) and so on.

The indispensable role of homochirality and chiral purity in life ([4](17–18))

The inherent relationship between life and molecular asymmetry has been appreciated since the time of Pasteur, almost 150 years ago. Today it is common knowledge that the essential biopolymers of life are characterized both by a unique homochirality as well as by the absolute enantiomeric homogeneity of their monomer units. Thus in protein polymers we find monomers consisting of only L-amino acids, with no D enantiomers present, while the nucleic acid polymers, RNA and DNA, accommodate only D-ribose and 2-doexy-D-ribose, respectively, with no L enantiomers in their mononucleotide monomer units. Why are homochirality and enantiomeric purity so obviously essential in these biopolymers? Two principal attributes of life, the synthesis of specific proteins mediated by DNA and RNA, and the replication of DNAs to transfer genetic information, depend for their functional efficacy upon the precise fitting together (complementarity) of the helical polymer strands involved in expediting the processes. Twenty-eight years ago Miller and Orgel [8] first pointed out that, although replicating double-stranded nucleic acid helices are possible with pure D- or pure L-ribotides, it is not possible to build such helices with mixtures of the two. Avetisov and Goldanskii [9] have illustrated this with molecular models, where they showed that both strands of a double-stranded matrix-replica structure can complement each other when both strands contain homochiral monomer units (e.g., all D-nucleotides). However, if one of the strands contains a monomer of opposite chirality, the complementarity is precluded since, for geometrical reasons which impede critical H-bonding between the strands, "the replica completely loses the matrix profile in the vicinity of the chiral defect" [9]. These conclusions follow also from earlier experimental studies of Joyce and co-workers [10], who showed that in template-directed oligomerizations of nucleotides, the construction of a complementary strand on a chirally pure template was strongly hampered if the monomers being incorporated into the new strand were not themselves also chirally pure. We have recently summarized [5] additional examples documenting the importance of chiral purity on the efficiency of oligomerizations leading to polypeptides and other polymers. Today it is generally accepted that homochirality and chiral homogeneity are prerequisities for the efficient self-replication of biopolymers, and thus also for the origin of life [11]. In short: No Homochirality, No Life. Since ordinary syntheses of chiral compounds result in racemic products, the question of how molecules of a single chirality arose in nature has been paramount since the time of Pasteur, and has led to many subsequent theoretical and experimental investigations. Before reviewing some of these, let us turn our attention briefly to the terrestrial environmental setting under which prebiotic chiral molecules would have had to originate.

Environmental and temporal constraints on prebiotic chirality ([4](41–42); [5](97–98))

An ever-present peril to chiral molecules in any environment is, of course, racemization. It has been estimated [12] that the racemization half-lives for amino acids is 10^5–10^6 years at ambient temperatures, and racemization rates are known to be enhanced by metal ions, alkaline conditions and ionizing radiation [13]. Thus chiral amino acids (and other prebiotic monomers) must have endured a precarious existence indeed in the aqueous prebiotic environment, where processes leading to and maintaining enantiomeric purity clearly "must have arisen in the course of chemical or biological evolution within a much shorter time interval than the racemization rate" [14]. Racemization has been deemed so hazardous, in fact, as even to have led Miller and Orgel in 1974 to the currently untenable conclusion that life must have originated from racemic precursors [8]. Nevertheless, an even greater menace to prebiotic chirality lurked in the chaotic and turbulent nature of the primitive environment itself. The earlier paradigms of Darwin's "warm little pond" and the gradually evolving, optically inactive [8] "primordial soup" of the Oparin-Haldane hypothesis [15] have recently undergone substantial changes [16]. It is now understood that during its first 700 Myr or so the primeval Earth was subjected to intense bombardment by innumerable comets and asteroids as large or larger than comet Halley, some of which impacted with sufficient energy to vaporize the oceans, form of rock-vapor atmosphere and sterilize Earth's surface to depths of tens of meters [17]. Such catastrophic impacts are thought to have frustrated the origin of life, perhaps repeatedly, during this period [18], and only after impact frequencies had diminished by orders of magnitude could life emerge and proliferate about 3.8 Gyr ago [19].

Over 20 years ago Sagan suggested [20] that such conditions could curtail the period available for the origin of life to "only a few hundred million years – perhaps much less", and it has been estimated more recently that the *maximum* time needed for the origin of life was as little as 2.5 to 133 Myr, depending upon whether $^{12}C/^{13}C$ ratios in sedimentary rocks or the oldest known microfossils are used as criteria for the earliest life [21]. Now, if the above environmental and temporal constraints on the origin of life are valid, and if replicating biopolymers are a prerequisite for the existence of life, then how much shorter must have been the actual time period accessible for the development of the molecular homochirality and enantiomeric homogeneity essential for the emergency of such biopolymers?

We turn next to a brief consideration of recent developments in some of the historically proposed suggestions for the terrestrial origin of molecular chirality, evaluating their potential efficacy in the hostile prebiotic environment described above. Such proposals fall into two broad categories: *random mechanisms*, which have an equal probability of generating either of the two possible enantiomers (e.e. = 0.0%), and *determinate mechanisms*, which produce a preponderance of molecules having a predetermined chirality (e.e. > 0.0%).

Random terrestrial mechanisms

Spontaneous symmetry breaking – autocatalysis ([4](19–22); [5](62–65); [6](5–10); [7](180–186))

The earlier theoretical contentions of Avetisov and Goldanskii that the prebiotic, racemic "organic medium" underwent a spontaneous and total symmetry breaking (deracemization) process prior to subsequent molecular evolution into replicating, information-carrying biomolecules have recently been eloquently reiterated [9]. In brief, they postulate a racemic monomer medium which somehow spontaneously converts into a state of chiral purity under the influence of "natural mechanisms of strong violation of mirror symmetry of geochemical or cosmochemical areas". This provides a source of homochiral monomers which then undergo "polymer takeover" to afford homochiral polymers now capable of developing into "informational and functional carriers at the biochemical level of complexity". These vivid argments, which have never received and are apparently immune to experimental appraisal, unfortunately fail to specify either the chemical nature of the "organic medium" or the mechanisms by which its mirror symmetry is broken, and are completely oblivious to prebiotic environmental constraints such as those considered above.

Accompanying their description of the self-assembly of higher oligomers of pyranosyl-RNA by ligative oligomerization of tetranucleotide-2′,3′-cyclophosphates, Eschenmoser et al. [22] have recently offered a novel and convincing mechanism for spontaneous symmetry breaking. Briefly, they suggest that, starting with a racemic mixture containing all possible diastereomers, sets of such "tetramers with different but mutually fitting base sequences can be expected to co-oligomerize stochastically and generate sequence libraries consisting predominantly of homochiral (D)- and (L)-oligomers". They then show how, after reaching a sufficient chain length, the oligomers are "driven beyond a critical level of constitutional complexity", such that a true racemic mixture of the oligomers is no longer possible. Then, with "internal and external selection pressures" either the homochiral (D)- or (L)-oligomer must eventually predominate. Although these novel suggestions are extrapolated from the authors' solid laboratory observations, they have yet to be demonstrated experimentally, and the system involved seems a trifle sophisticated chemically to be revelant in the tempestuous prebiotic environment.

In 1953 Frank developed a mathematical model for the "autocatalytic" random symmetry breaking of a racemic system [23]. In this model the reaction of one enantiomer yields a product which acts as a catalyst for the further production of more of itself, and an inhibitor for the production of its enantiomer. He demonstrated that such a system is kinetically unstable, such that any random fluctuation producing a transient e.e. in the 50:50 population of the racemic reactant would result in the eventual domination of the product from that reactant in transient excess. In 1959, unaware of Frank's paper, Calvin proposed [24] a similar scheme, "stereospecific autocatalysis," for autocatalytic symmetry breaking. Calvin's meachnism has been validated experimentally in the context of the total spontaneous resolution during the crystallization of certain *conglomerates* (racemic mixtures containing equal numbers of separate crystals of each enantiomer). In the crystallization of a conglomerate, either the crystals of one enantiomer or the other may spontaneously separate, leaving the other enantiomer in solution. If there now is the possibility of the equilibration of the enantiomers in solution, and if the enantiomer in solution can convert rapidly to the enantiomer which is crystallizing before crystallization is complete, then the entire racemate may deposit as a single enantiomer. At least half a dozen examples of Calvin's stereospecific autocatalysis involving the *spontaneous resolution under racemizing conditions* (SRURC) of conglomerates have substantiated the mechanism. SRURC has a serious flaw, however, for realistic prebiotic symmetry breaking. Even if a hypothetical conglomerate were to undergo total spontaneous resolution by crystallization, for example, from some evaporating tidepool on the primitive Earth it would, barring subseqent implausible solid-state reactions, inevitably racemize on redissolving in the aqueous environment prior to undergoing the next stage of its prebiotic chemical evolution. In the 45 years since its proposal, Frank's mechanism has spawned innumerable theoretical refinements [25–26], including consideration of such factors as reversibility, racemization, environmental "noise" and parity-violating energy differences. In contrast to the above examples of stereospecific autocatalysis by the SRURC of conglomerates, however, none of these theoretical suggestions is supported by even an inkling of experimental evidence. While earlier attempts to validate the Frank mechanism for the autocatalytic amplification of small e.e.s in other experimental systems have generally been unsuccessful, several recent attempts have shown more promising results [27–29]. Both these experiments and those described earlier validating the mechanism of stereospecific autocatalysis, however, have all involved exotic chemical systems which, while they are of interest in connection with the Frank hypothesis, are entirely implausible in any realistic prebiotic environment.

Solid-state asymmetric syntheses in chiral crystals ([4](24); [5](65); [6](10–16))

Certain nonchiral organic molecules have been observed to crystallize randomly in enantiomorphic crystal forms. When such an individual chiral crystal reacts in the solid state with itself or with another nonchiral reactant intimately incorporated into the chiral crystal lattice, an optically active product may result. Here the particular chirality of the enantiomorphic crystal imposes a unique chirality on the product of the solid-state reaction occurring within the chiral crystal lattice. A number of such lattice-controlled solid-state asymmetric syntheses, as well as closely related reactions, have been described in the past [30], and many of the authors have porclaimed such processes as potentially accountable for the origin of chiral molcules in nature. Unfortunately, these experiments have necessitated contrived reactants far removed from biological systems, as well as delicate and carefully controlled reaction conditions, restrictions which clearly render such mechanisms completely implausible in the actual primordial environment.

Asymmetric processes involving quartz and clay minerals ([4](22–24); [5](66–68); [6](16–22), [7](192–204)

The silica mineral *quartz* is found in nature as either right-, (+)-, or left-handed, (–)-, morphologically chiral crystals. In 1935 Japanese workers reported that certain racemic cobalt complexes in solution could be resolved by *asymmetric adsorption* on quartz crystals, with (+)-quartz preferentially adsorbing one enantiomer of the complex and (–)-quartz of the other enantiomer. Three years later such observations were repeated by investigators in Athens, and it was first suggested that asymmetric adsorption on quartz might have engendered the first optically active compounds in nature. A number of early workers also reported asymmetric catalytic capabilities of chiral quarts crystals, either alone or when coated with thin layers of certain metals, and made similar suggestions. Extremely careful investigations in 1968 by French scientists, however, showed that such claims were based uniformly on artifacts, and the general validity of such phenomena became questioned. Our own work and that of others in the 1970s, however, subsequently demonstrated that asymmetric adsorption on finely powdered quartz could indeed occur if caried out under carefully controlled conditions using anhydrous solvents. These stringent restrictions, however, plus the fact that the terrestrial distribution of (+)- and (–)-quartz is completely random (see below) render it quite inconceivable that any process involving chiral quartz crystals could have been implicated in the prebiotic origin of molecular chirality.

Although the clay minerals kaolinite and montmorillonite have no chirality associated with their crystal structures, nevertheless both the asymmetric polymerization of amino acids and the asymmetric adsorption of both amino acids and sugars have been attributed to these clays. Our own and other later investigations, however, disclosed that such reports were again based on artifacts, and to date there is no experimental or theoretical evidence whatsoever for asymmetric interactions between clay minerals and chiral organic molecules.

Determinate terrestrial mechanisms

Electric, magnetic and gravitational fields ([4](25); [5](69–72); [6](54–58))

In the mid- 1800s Pasteur explored unsuccessfully the possibility that magnetic or gravitational fields might induce asymmetric effects in chemical reactions. His failures were explained by Curie 40 or so years later as due to the fact that the inducing fields were themselves not chiral. Despite this, from 1939 to the mid- 1990s number of investigators have reported asymmetric syntheses involving electric, magnetic and gravitational fields, frequently suggesting their potentiality for the origin of chirality. Certain of these claims have been challenged on theoretical grounds, others have been shown to be due to experimental optical rotation errors or other artifacts, while still others, since corrected, have been disclosed as intentionally fraudulent. To date there are no experimentally validated clains for asymmetric syntheses resulting from the influence of such fields, although the possibility of magnetically induced chiral effects under specific and carefully contrived experimental conditions still remains an open theoretical question [31].

Circularly polarized light ([4](25–27); [5](72–76); [6](58–69); [7](204–210))

Circularly polarized light (CPL) is a light wave whose electric vector spirals clockwise to the right (RCPL) or counterclockwise to the left (LCPL) along the axis of propagation. R- and LCPL represent physical agents characterized by "true chirality" [32], and are capable of interacting stereoselectively with chiral or prochiral molecules in photochemically induced reactions [33]. Over a century ago it was suggested that CPL might be implicated in the origin of chiral molecules in nature, and in 1929–30 the first experimental demonstrations of this possibility were provided with the successful asymmetric photolyses of racemic ethyl α-bromopropionate, $CH_3C^*H(Br)COOEt$, and racemic *N,N*-dimethyl-α-azidopropionamide, $CH_3C^*H(N_3)CONMe_2$, with ultraviolet R- and LCPL. The numerous successful CPL-induced reactions which have been reported since then fall into three categories: (i) *Partial photoresolution*, where a racemate is photochemically equilibrated to a mixture of enantiomers having e.e. > 0; (ii) *asymmetric photochemical synthesis*, where a chiral product is formed photochemically from a prochiral reactant; and (iii) *asymmetric photolysis*, where a residue having e.e. > 0 results after the partial photodegradation of a racemate. The first two processes could not be important prebiotically since they are limited (so far) to exotic and biologically irrelevant substrates, and since they are restricted theoretically to eliciting only trivial e.e.s ($\leq 1\%$). In asymmetric photolysis the two enantiomers are decomposed at different rates with either R- or LCPL, such that interruption of the photolysis before completion results in an unphotolyzed residue containing an excess of the more photostable enantiomer. By contrast, the prebiotic potential of asymmetric photolysis results from its general applicability to any racemic substrate having absorption bands at visible or ultraviolet wavelenths, and to the fact that that e.e.s up to very high values are possible theoretically and have been observed experimentally.

As to the availability of terrestrial RCPL or LCPL in the prebiotic environment, a number of sources ultimately dependent on sunlight have been suggested theoretically and/or observed experimentally. However, they all have the common drawbacks of providing both low intensities and only trivial excesses of a given CPL-handedness, and of being subject to both temporal and geographical randomization. Thus even the efficient and general mechanism of asymmetric photolysis could hardly be of terrestrial importance on the primitive Earth. The extraterrestrial application of this mechanism will be considered below.

Mechanism based on parity violation

In simplified terms the parity principle maintains that natural processes are invariant under spatial reflection, that is, they are also valid as seen in a mirror reflection. In 1956 Lee and Yang made their Nobel prize-winning prediction that the parity principle should be violated for certain "weak interactions", such as those involved in the β-decay of certain radioactive nuclides. This conjecture, verified experimentally a year later, has been the impetus in the intervening 40 years for countless theoretical and experimental efforts to connect such parity violation at the elementary particle level with biomolecular homochirality which, as seen above, is clearly a violation of parity at the biochemical level.

β-Decay effects ([4](27–33); [5](77–84); [6](24–48); [7](210–215))

The first attempt to make this connection was that of Vester and Ulbricht in 1959, who pointed out that when the longitudinally polarized "left-handed" electrons emitted with parity violation during β-decay impinge upon matter they generate circularly polarized "Bremsstrahlung" photons, and that these in turn might interact with prochiral chemical systems to yield chiral products by stereoselective CPL mechanisms such as those described above. They thereupon conducted numerous unsuccessful synthetic and degradative experiments, using a number of β-emitting nuclides to irradiate various organic substrates, and finally concluded that more powerful radioactive sources and longer irradiation times would be required to generate observable optical activity. Numerous and diverse experiments meeting these criteria were subseqently conducted by ourselves and by others in efforts to validate the Vester-Ulbricht hypothesis. These experiments generally involved the attempted asymmetric degradation of racemic substrates with the β-radiation and/or Bremsstrahlen from such β-emitters as ^{90}Sr, ^{90}Y, ^{152}Eu, ^{108}Ag, ^{3}H, ^{14}C and ^{32}P. However, in no cases were reproducible stereoselective effects observed.

Supplementing the above experiments using β-decay electrons and Bremsstrahlen from radionuclides, a number of experiments, again by ourselves and other investigators, have explored the possibility of *direct* asymmetric interactions between racemic substrates and a variety of artificially produced spin-polarized chiral particles, including electrons, protons, positrons and muons. Again, however, no reproducible enantioselective effects beyond experimental error have been reported.

In many of the above studies a new phenomenon termed *radioracemization* has been reported [13]. In these experiments, involving the incomplete radiolysis of an enantiomerically enriched substrate by ionizing radiation, partial racemization of the enriched enantiomer remaining in the undestroyed substrate was noted. It was emphasized that radioracemization might diminish, but would by no means necessarily negate, any stereoselective effects which might be caused by the action of spin-polarized chiral particles on racemic substrates.

Parity violating energy differences ([4](33–37); [5](84–89); [6](48–54)

The mirror image structures of D(–)- and L(+)-alanine shown in Figure 1 do not represent true enantiomers. To be genuinely enantiomeric in every sense, one enantiomer of a pair must not only be the mirror image of the other, but must also be composed of antimatter. Since enantiomers are thus not strictly degenerate, this represents a parity violation between any two enantiomers in the real world, which in turn results in a small energy difference (PVED) between them. This was first predicted in 1966 by Yamagata [34], who also suggested that PVEDs should engender small differences in the physical and chemical properties of enantiomers which, if amplified by an "accumulation principle" during polymerizations, could result in the homochiral biopolymers characteristic of our biosphere. The magnitude of PVEDs became the subject of several theoretical investigations during the 1970s, and from the early 1980s to the present a plethora of *ab initio* calculations by Mason, Tranter and MacDermott [2] involving the PVEDs of amino acids, proteins, D-sugars and derivatives, as well as DNA and quartz have dominated the field. The cumulative conclusions of such calculations are that the equilibrium populations of "natural" enantiomers, L-amino acids (free and in proteins), several D-sugars, and (–)-quartz) are favored over the corresponding "unnatural" enantiomers to the extent of $\sim 10^{-17}$. Put into a temporal perspective, 1 part in 10^{17} corresponds to approximately 1 s out of the 4.5 billion years since our Solar System was formed! This minuscule enantiomeric abundance of the "natural" enantiomers was then alleged to be capable of amplification into a state of enantiomeric homogeneity within the geologically instantaneous time span of only 10,000 years by a mechanism proposed by Kondepudi and Nelson in 1983 [35], which was itself a formalization of the Frank "autocatalysis" scenario [23] discussed above. Other proposed mechanisms for amplifying trivial e.e.s due to PVEDs into enantiomeric purity have been related to Yamagata's accumulation principle [34] and to Salam's "phase transition" (see below). It should be emphasized that all such conclusions are completely speculative and are unsupported by a single parcel of experimental evidence, and that theoretical reservations as to the efficacy of PVEDs to engender chiral molecules were advanced early on by Keszthelyi, Rein, Hegstrom, Morozov and others.

Persistent and still reiterated [2] allegations that a PVED is responsible for an alleged 1.0% terrestrial excess of (–)-quartz (i.e., 50.5% (–)- and 49.5% (+)-quartz) are based on a 1962 survey of 16,807 quartz samples, and repeatedly ignore a much larger 1978 survey of 27,053 samples which showed 50.17% (+)-quartz and 49.83% (–)-quartz. Thus the terrestrial distribution of quartz enantiomers cannot be cited for authentication of PVED effects.

As discussed above, the only autocatalytic amplification mechanism that has been validated experimentally has been the "spontaneous resolution under racemizing conditions" (SRURC) during the crystallization of certain conglomerates, and such processes have been shown to be capable of developing small e.e.s into larger ones, even approaching 100%. Thus in principle SRURC should be capable of amplifying the 10^{-17} e.e.s produced by PVEDs into larger observable ones, providing only that the PVED effects are greater than the random noise of statistical fluctuations. This possibility has been investigated on a statistical basis in two separate SRURC systems, the crystallization of 1,1′-binaphthyl and derviatives studied by Pincock and co-workers in the early 1970s, and of 1,4-benzodiazepinooxazole derivatives studied by Okada and co-workers in the late 1980s and by Bonner in the 1990s [36]. In neither case was any enantiomeric bias observed during repeated crystallizations. The completely random nature of the SRURC data thus gives no support for the capability of PVEDs to engender observable net chirality. Sustaining these conclusions are recent observations of Kondepudi and co-workers [37], who confirmed that crystallizations of sodium chlorate lead to the random formation of optically active crystals, whether the crystallization were from undisturbed solutions (producing equal numbers of (+)- and (−)-crystals) or from stirred solutions (producing essentially equal numbers of groups of either all-(+) or all-(−) crystals).

The most recent proposal for global symmetry breaking due to the PVED is that of Salam [38] in 1991, who postulated that a "Bose condensation" phase change from the less stable enantiomer of a pair into the more stable one should occur spontaneously below some critical low temperature, leading eventually to enantiomeric homogeneity. However, subsequent meticulous experiments by Figureau and co-workers in 1995 to substantiate this claim, in which DL-cystine was subjected to temperatures as low as 0.01 K, have failed to detect any evidence for the predicted phase transition into L-cystine [39].

Evaluating the possibilities of generating molecular chirality by mechanisms based on parity violation, whether through the auspices of β-decay or PVED effects, we must conclude that there is neither an uncontested theoretical consensus nor reproducible experimental evidence to support the existence of any causal relationship whatsoever between parity violation at the elementary particle level and that at the biochemical leve i.e., the unique homochiralities of biomolecules.

Extraterrestrial sources of homochirality ([4](42–44))

We have seen above that both random and determinate terrestrial mechanisms historically proposed to explain the prebiotic origin of chiral molecules have been either unsuccessful, irrelevant or realistically implausible in the violent environment characteristic of the primitive Earth. Now, if there exists no viable terrestrial source to supply the chiral precursors necessary for the development of homochiral biopolymers, the only logical alternative is an extraterrestrial source. We turn now to a consideration of this alternative.

Based on the 1953 experiments of Miller and Urey in which electric sparks (simulating lightning) in gaseous mixtures of methane, ammonia, hydrogen and water vapor (simulating the atmosphere), as well as on numerous similar experiments by others which also produced complex mixtures of organic compounds, formation of the organic precursors in the "primordial soup" of the Oparin-Haldane hypothesis became generally attributed to the interaction of various indigenous energy sources with the presumed reducing atmosphere of the primitive Earth [15]. A supplemental premise was advanced by Oró in 1961 [40], who first suggested that the precursors of life's biomolecules might also have been formed extraterrestrially, incorporated into comets and then transported to Earth by cometary impact. Today it is generally acknowledged that extraterrestrial sources, particularly comets [41–44] and interstellar dust grains [45–46] provided a continuing source of both simple and more complex prebiotic organic precursors to the primitive Earth, and estimates of the realtive importance of such exogenous extraterrestrial as compared to the above endogenous terrestrial sources of been attempted [47]. How does this newer paradigm accommodate chiral molecules?

In 1980 Khasanov and Gladyshev first suggested that chiral molecules might also have originated in outer space under the influence of some unspecified "stereospecific effect", and three years later Rubenstein and co-workers [48] outlined a specific mechanism which would provide an extraterrestrial source of chiral molecules, a scenario which involved the circularly polarized synchrotron radiation photons originating from the neutron star remnants of supernovae. We shall now consider the background and details of this hypothesis briefly, and review recent experiments and observations which tend to substantiate it.

Stars are maintained by energy-releasing nuclear fusion reactions which occur as the star collapses gravitationally and heats up: H $\rightarrow$ He (in small stars or the outer layers of large stars); He $\rightarrow$ C; C $\rightarrow$ Ne,O,Si; and Si $\rightarrow$ Fe (at the core of large stars). When a large star of 8 to 10 or more solar masses has depleted all of its nuclear "fuel", heat production by fusion at the core stops, as further nuclear transformations beyond Fe consume rather than release energy. Since the energy released no longer balances the gravitational attraction of the star's mass, the star collapses gravitationally "as one piece" in a matter of seconds to an incompressible core of nuclear density. This generates a rebounding shock wave which turns the stellar collapse into a colossal *supernova* explosion which blows off the outer mantle of the star, producing ejecta of refractory interstellar dust, leaving a *neutron star* core, and releasing prodigious amounts of energy, which in turn compress the interstellar medium to form new star systems (such as our solar system). Supernovae are by no means rare events, three having been recorded in our galaxy within the past 1000 years. At this frequency some 14 million supernovae have detonated in our own galaxy in the 4.7 billion years since our solar system itself was formed by a supernova. Since they are intrinsically so bright, 10 or more supernovae are sighted each year in distant galaxies [49].

The neutron star resulting from a supernova is thought to be a tiny (~10 km diameter), incredibly dense (~10^8 g/cm^3) core possessing a prodigious magnetic field (~10^{13} G, as compared to 0.5 G for Earth), emitting and surrounded by a plasma of electrons and protons and, in conserving the angular momentum of the original star, rotating at tremendous velocities (milliseconds per revolution, ~ 10^4 r.p.m.). The rapidly rotating magnetic field around the core creates colossal electric fields which in turn accelerate the charged particles in the plasma around the core to near relativistic velocities. However, constrained by the intense magnetic field to move in circular orbits, the accelerating electrons around the core thereupon emit synchrotron radiation, as originally suggested by Shlovskii for the spectrally continuous light emitted from the Crab nebula (the remnant of a supernova which occurred on 4 July, 1054 AD [50]. Synchrotron radiation is polychromatic and known to be plane-polarized as emitted in the plane of the orbiting electrons, but elliptically and ultimately circularly polarized with one handedness above and the opposite handedness below the plane of orbit [51]. In the vicinity of a neutron star this situation results in vast regions of space above and below the "equator" of the star which are continuously irradiated with elliptically and circularly polarized synchrotron photons, including CPL of all photochemically relevant wavelengths.

Fifteen yers ago Greenberg proposed a detailed hypotheses for the formation of organic mantles on dust grains in dense interstellar "molecular clouds" [52]. He suggested that these submicroscopic silicate dust particles become covered with "ices" formed by the freezing on their surfaces of small interstellar molecules such as H_2O, NH_3, CH_4, CO, CO_2, CH_2O and so on present in the clouds. The small molecules in these ices then undergo photolysis by ultraviolet radiation into reactive free radicals, which remain trapped in the ices. On subsequet slight warming the radicals recombine randomly to form more complex organic molecules, up to high molecular weight polymers, which are then left as residues on the grains as the ices evaporate. Repetitions of such an accretion-photolysis-recombination sequence of events ultimately produce interstellar grains coated with thick outer mantles containing a wide variety of increasingly complex organic molecules. Greenberg and co-workers have supported this hypothesis with laboratory experiments wherein mixtures of such interstellar gases were frozen as ices in high vacuum on a cold (10 K) surface simulating an interstellar grain, then were irradiated with ultraviolet light. Investigation of the residues produced on the surface by gas chromatography and mass spectrometry revealed a host of organic compounds, some of which were racemates of potentially optically active molecules [53].

In 1987 we extended the suggestions of Rubenstein and co-workers [48], combining the above facts and theories into the unified hypothesis for the extraterrestrial formation of chiral molecules and their transport to Earth illustrated in Figure 2 [54]. Here we postulate that the ultraviolet CPL components of the synchrotron radiation illuminating the immense domains of space "above" or "below" the neutron star remnant of a supernova stereoselectively photolyze the racemic constituents of the organic mantles on interstellar grains by the familiar and efficient mechanism of partial asymmetric photolysis, forming chiral compounds having a slight excess of one enantiomer in the grain mantles. The enantiomeric integrity of the optically active compounds thus produced in the mantles would be protected indefinitely against racemization by the cold interstellar temperatures (≤10 K).

Greenberg has recently delineated four conditions which must be satisfied if the neutron star hypothesis in Figure 2 is to be valid [55]: (i) pre-

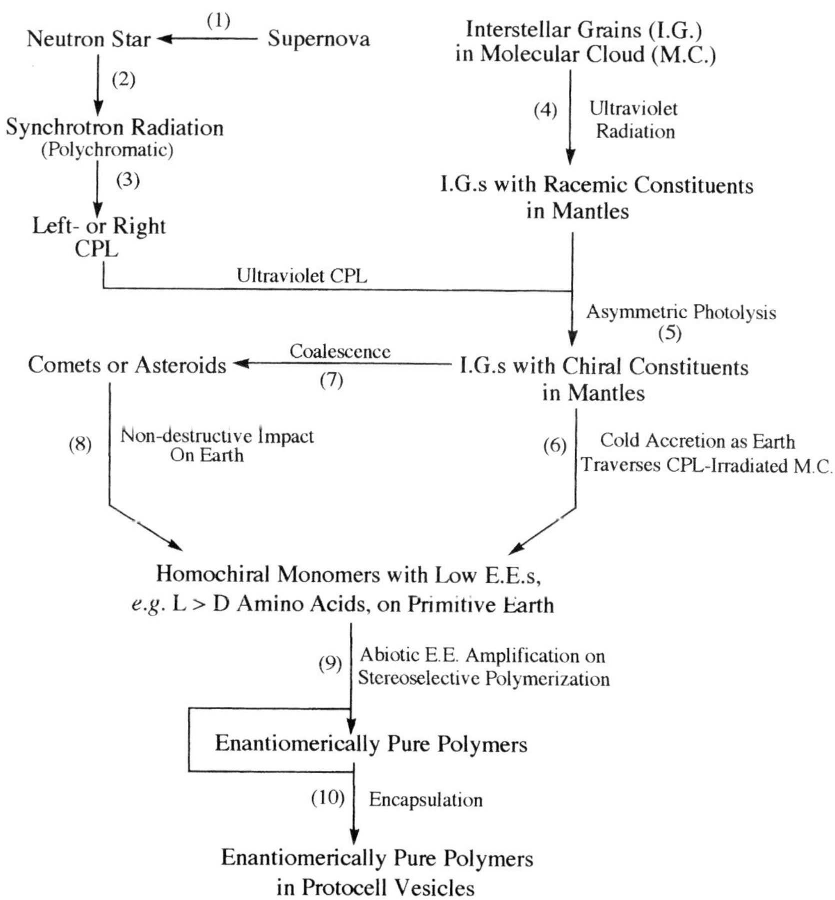

Figure 2. Extraterrestrial source of homochiral polymers.

biotic molecules with the potential for chirality must exist on interstellar grain mantles; (ii) CPL must induce asymmetric photolysis of racemates at interstellar temperatures just as it does at room temperature; (iii) there must be a sufficient number of neutron stars with suitable ultraviolet radiation intensity to have created significant e.e.s in the mantles of grains in individual molecular clouds; and (iv) the grains with their mantles of chiral molecules must be transportable to Earth's surface in such a way as to preserve the e.e.s in the constituents of the mantles. Let us consider each of these conditions in greater detail.

Condition (i) involving potentially chiral molecules on interstellar grains has already been validated by the experiments described by Briggs and coworkers, where such possibly chiral molecules as glyceric acid, glyceramide, alanine and serine, as well as glycerol, a potential substitute for ribose in simpler pre-RNA nucleic acid polymers, have been produced by the ultraviolet irradiation of frozen interstellar gas mixtures as described above [53, 55]. Condition (ii), requiring asymmetric photolysis at interstellar temperatures, has also been investigated experimentally [55]. Using a cryogenic irradiation apparatus as previously described [53], Greenberg exposed thin layers of racemic tryptophan at 10 K to 252.4 nm ultraviolet CPL for up to 100 h, then analyzed in replicate the partially photolyzed residues using high-pressure liquid chromatography (HPLC) to determine the enantiomeric composition of the remaining tryptophan. Considering the thickness of the irradiated samples and the ultraviolet CPL penetration of the samples, Greenberg estimates that D > L e.e.s as high as 50% were formed in the tryptophan which was actually photolyzed. Thus the asymmetric photolysis of racemates proved indeed to be a workable mechanism at interstellar temperatures. Condition (iii), having to do with the number and ultraviolet radiation efficiency of neutron stars, has also been addressed in detail by Greenberg, using the Crab nebula pulsar as a "standard" neutron star [55]. By calculating the mean density of "effective" neutron stars in our galaxy and the mean cross-section for collisions of these with randomly moving interstellar clouds, he concludes that the time between star-cloud collisions is $\sim 3 \times 10^8$ years. Since the mean lifetime of a dust cloud is $\sim 3 \times 10^7$ years, he deduced that about 10% of the molecular clouds which become solar systems have dust grains with mantle components having e.e.s of up to 50%, and that all presolar molecular clouds would have e.e.s of ~10% in their grain mantles. It thus appears that a large fraction of the solar systems in our galaxy would be formed from molecular clouds which had significant e.e.s in their dust grain mantles [56]. As to condition (iv), regarding the accretion of Earth of interstellar grains with the chiral components in their mantles intact, let us next consider several possibilities which exist for its implementation.

As our solar system revolves around the center of our galaxy at 110-million-year intervals, it intersects with dust clouds in the spiral arms of the galaxy. Earth has thus traversed such galactic dust domains 40 or so times since the solar system was formed, and 5 or 6 times prior to the earliest life on Earth, ~3.8 Gyr ago, Greenberg estiamtes that at each traversal Earth accretes some 10^8-10^{10} metric tons of matter, far more than the total current biosphere. He argues further that organic molecules in the mantles of the dust grains would be transferred to Earth unpyrolized due to the small size of the grains, and that such cold accretion would have provided some of the complex organic molecules which preceded life on Earth [52].

Interstellar grains are also thought capable of aggregating into clumps, and these in turn into larger clumps, and so on to comet dimensions, such that a comet nucleus is primarily a fluffy aggregate of submicron grains [43]. It has been estimated that one comet the size of Halley carries as much as 10% of Earth's biomass in the form of refractory organics at its aggregated dust core [53]. Thus a relatively few comets impacting Earth toward the end of the major comet impact epoch could have provided not only the oceans [57] but also a large fraction of the prebiotic molecules which were the precursors of life [58]. To circumvent the severe heating and pyrolysis of these prebiotic organics which would occur during impact of a comet with Earth, Greenberg and coworkers emphasize that several important conditions must prevail [59]: (i) the comet nucleus must have a low mean density, such that it may be adequately decelerated and cushioned by a dense atmosphere during entry; (ii) the nucleus must disintegrate into smaller fragments during entry; and (iii) the ice mantles on the dust grains in these smaller fragments must act as heat shields, which on evaporation allow the underlying grains with their organic mantles to float gently to Earth. In contrast to the random and uniform global accumulation of interstellar dust *per se*, such a cometary impact scenario has the advantage of permitting possible high aqueous concentrations of chiral prebiotic molecules in limited locations, each reminiscent of the "primeval procreative comet pond" postulated by Clark [60].

Recently ultraviolet and visible wavelength "visible imaging system" cameras aboard NASA's 1996 "Polar" satellite have confirmed [61] the earlier controversial detection of 25–40 km diameter "atmospheric holes", which appeared as dark spots in photographs taken by L.A. Frank and co-workers of the University of Iowa while imaging the Earth [62]. These "holes", which occur at altitudes of 2000 to 3000 km, have been shown to move in a general east-to-west direction over Earth and to be characterized by OH emission due to the photodissociation of water. Frank and his co-workers have suggested that the "holes" are caused by a new class of objects in the Solar System, small, fluffy cometlike water-ice "snowballs" weighing ~100 tons. These enter the upper atmosphere at rates of some 5–30 per minute, whereupon they disintegrate at altitudes of 600 to 15,000 miles and gently rain down on Earth, providing a global mass accretion of ~10^{12} kg/y [62]. Frank has proposed that this "relatively gentle 'cosmic rain' – which possibly contains simple organic compounds – may well have nurtured the development of life on our planet" [63]. It should be emphasized that Frank's hypothesis of the repetitive infall of "miniature comets" has been from the outset [64] and still remains [61] the subject of intense controversy. Finally, moreover, if Frank's accretion mechanism were indeed to prove valid, it would suffer the same shortcoming as that involving the direct accumulation of interstellar dust, namely a uniform and extremely dilute dispersal of chiral molecules on the surface of the primitive Earth.

While confirmation of the suggestion that comets may have conveyed extraterrestrial chiral organic molecules to the prebiotic Earth must probably await direct cometary exploration with space probes, the proposition is strongly supported by recent analyses of some of the amino acid constituents of the Murchison meteorite, a carbonaceous chondrite which fell to Earth in fragments near Murchison, Victoria, Australia, in late September 1969. Many of the fragments, which may contain up to 3% by weight of carbon, were collected in a "pristine state" and immediately subjected to extraction and analysis of their organic constituents, before indiscriminate terrestrial contamination could occur. While a monumental array of simple to complex organic compounds has been found in the Murchison and similar carbonacous chondrites [65], two recent findings regarding amino acids from the Murchison meteorite are particularly relevant to the neutron star hypothesis.

First, while earlier studies of the enantiomeric compositions of the amino acids in the Murchison stone reported them to be essentially racemic, thus precluding terrestrial contamination with L isomers and suggesting an extraterrestrial origin [66], Engel and co-workers [67, 68] have more recently disclosed that its L-amino acids exceed its D-amino acids by as much as (for alanine) 18%. Moreover, measurements of the stable isotope ratios, $^{13}C/^{12}C$ [68], D/H and $^{15}N/^{14}N$ [69], for the amino acid constituents of the Murchison meteorite indicated that these constituents were unambiguously of extraterrestrial origin, "suggesting that optically active materials were present in the early Solar System before life began" [68]. Second, in 1997 Cronin and Pizzarello reported that gas chromatography-mass spectral analyses for enantiomeric ratios in two nonprotein amino acids from the Murchison meteorite, isovaline and α-methylnorvaline, showed an excess of L over D enantiomers of 9.2 and 2.8%, respectively [70]. Since α-methylnorvaline has never been found terrestrially and isovaline has been reported only in certain fungal antibiotics, terrestrial contamination is again precluded. Thus the results of both the Engel and Cronin studies clearly indicate an extraterrestrial origin for the significant excesses of L over D enantiomers among the Murchison amino acids. Moreover, since these amino acids have been subjected to racemization by liquid water and heating in the parent asteroid during the ~4.5 Gyr since its formation, as well as to radioracemization by cosmic rays and indigenous radioactivity [71], the e. e. s reported above clearly represent lower limits of L-amino acid enrichment.

While the L-enantiomeric excesses reported for the Murchison amino acids in the above studies are clearly consistent with an supportive of the neutron star hypothesis summarized in Figure 2, the findings also allow one to make a number of both definitive and tentative additional inferences. First, e.e.s are capable of being generated in asteroids and presumably comets by an extraterrestrial mechanism(s). Second, such extraterrestrial e.e.s are capable of being transported to Earth without their total destruction. Third, e.e.s can accordingly accumulate on Earth in the absence of living organisms. Fourth, the existence of e.e.s in extraterrestrial bodies is not necessarily indicative of the presence of life in these bodies. Fifth, the search for extraterrestrial e.e.s as "signatures" of extraterrestrial life is misguided, since such e.e.s are merely a necessary but not a sufficient criterion of life. And more tentatively – sixth, the L homochirality of the optically active amino acids in the Murchison meteorite would imply a similar L homochirality of amino acids in other asteroids and comets in our Solar System. Seventh, the homochiral L-amino acids in these bodies could be transported to Earth without total racemization or destruction, thus providing a biased impetus for the L homochirality of amino acids in our biosphere.

Finally, the neutron star hypothesis in Figure 2 permits us to make two predictions, which unfortunately cannot be verified at present. First, because of the frequent occurrence of supernovae in our own and other galaxies and the prevalence of molecular clouds of "stardust" containing organic-rich mantles, life based on chiral organic molecules should be an inevitable and relatively frequent phenomenon in the Universe. Second, due to the overall symmetry of the RCPL and LCPL synchrotron radiation illuminating the regions "above" and "below" a neutron star, the homochirality of the biomolecules characteristic of extraterrestrial life forms need not necessarily correspond to that on Earth, but should vary randomly throughout the Universe.

The amplification and sequestration of molecular chirality

The various terrestrial and/or extraterrestrial determinate mechanisms discussed above which were in fact capable of generating molecular chirality proved generally to afford only small e.e.s, ranging from the ~20% reported in some asymmetric photolysis experiments to the 10^{-17} calculated theoretically for PVED effects. Whatever the mechanism, it is accordingly obvious that subsequent amplification processes must have been operative on the primitive Earth to develop such small e.e.s into the high degree of homochirality and enantiomeric purity necessary to permit the emergence of self-replicating biopolymers. For these reasons a number of both theoretical and experimental investigations have been undertaken in the past with the objective of discovering effective abiotic mechanisms for e.e. amplification. We now consider these briefly, again emphasizing their inherent plausibility in the austere and brutal prebiotic environment.

Amplification by autocatalysis ([4](38–39); [5](92–93); [6](73–75))

The various autocatalytic schemes for spontaneous symmetry breaking discussed above would, of course, also provide theoretial mechanisms for the amplification of small e.e.s. Accordingly, such schemes have been frequently suggested [2, 9, 35] as possible mechanisms to achieve the chirally pure "starting point" for prebiotic molecular evolution toward our present homochiral biopolymers. Such demonstrably successful amplification mechanisms as the spontaneous resolution of conglomerates under racemizing conditions [36], lattice-controlled solid-state asymmetric reactions [30] and other autocatalytic processes [28, 29], however, have uniformly lacked generality and have involved such specialized reactions and experimental conditions as to render them unrealistic and implausible on the primitive Earth.

Other experimentally successful mechanisms which have been proposed for chirality amplification, such as those involving "kinetic resolutions" [72], enantioselective occlusions of enantiomers on opposite crystal faces [73] and lyotropic liquid crystals [74], are interesting in themselves but can hardly be of prebiotic relevance, again owing to their limited scope and the specialized experimental conditions needed for their implementation. Mechanisms involving the partial evaporations of [75] or partial precipitations from [76] solutions in which there exist solubility differences between a racemate and its individual enantiomeric constituents have been studied in only a few systems. Such processes might be intuitively plausible in a primitive Earth setting, but their generality at the present level of knowledge unfortunately appears quite limited.

Amplification by polymerization ([4](39–41); [5](93–97); [6](75–84); [7](217–221))

In 1957 Wald first proposed [77] that the secondary α-helix structure of a polypeptide chain should bias the selection of amino acid enantiomers toward homochirality as polymerization progressed. In other words, the particular chirality of the α helix should dictate that of chirality of the new monomers adding to it be of the same handedness as those already making up the helix. This seminal suggestion soon received abundant verification in a variety of prototypical experiments involving base-catalyzed conversion of monomeric amino acid *N*-carboxy anhydrides (NCAs) into polypeptides. These lead us in the mid-1970s to initiate a series of model experiments designed to explore the further possibility that Wald's mechanism might accordingly be applicable for the amplification of the chirality of enantiomerically impure amino acids.

Our studies involved the partial polymerization of NCAs of mixtures of specific amino acids having known e.e.s., followed by determination of the e.e.s of the amino acids in both the resulting polypeptides and in the residual unreacted NCA monomers [78]. In a typical experiment we found that when an optically impure leucine NCA monomer having an L > D e.e. of 31.2% was polymerized to the extent of 52% to the helical polyleucine peptide, the e.e. of the polymer was enhanced to 45.4%, an increase of 14.2%, while the e.e. of the unreacted leucine NCA monomer was depleted to a similar extent. Analogous experiments with valine NCAs of known e.e.s. however, led to a reverse effect, namely the preferential incorporation of the racemate rather than one enantiomer into the growing polyvaline peptide. This was interpreted as due to the fact that polyvaline consists of β sheets rather than as α helices, emphasizing that the Wald mechanism applies only to α helix polymers. At about the same time Brack and Spach [79] showed that, under proper conditions, β sheet polymers could also be implicated in the amplification of amino acid e.e.s.

In our above experiments with polyleucines we also observed that the partial hydrolysis of mixtures of nonhomochiral leucine peptides led to the preferential hydrolysis of those components of the mixtures which were enantiomerically more random, leading to a corresponding e.e. enrichment in the unhydrolyzed polymer [80]. For example, when the above polyleucine sample having an L > D e.e. of 45.4% was 27% hydrolyzed, the residual unhydrolyzed polymer had an e.e. of 55%, representing an e.e. increase of 9.6%. From the two typical experiments cited we see that a sequential combination of partial polymerization, with its 14.2% e.e. increase, and partial hydrolysis, with its 9.6% e.e. increase, resulted in an impressive overall L > D e.e. enhancement of 23.8% in just two steps from the initial leucine NCA monomer to the final polyleucine peptide. The success of these model experiments led us to propose a mechanism involving repetitive cyclic sequences of partial polymerization – partial hydrolysis steps, driven by environmental dry and wet cycles, which could ultimately lead to homochiral polypeptides on the prebiotic Earth [81]. Brach and Spach [82] have also proposed a mechanism involving partial hydrolysis for the enantiomeric enrichment of polypeptides having β sheet secondary structures. While such model experiments involving e.e. amplification of amino acids during polymerization have admittedly involved prebiotically "unrealistic" substrates and carefully contrived experimental conditions, it is noteworthy that both secondary structures of proteins, α helices and β sheets, have been found capable of acting stereoselectively to provide for e.e. enhancements during such model polymerizations. Analogous amplification experiments embodying prebiotically realistic environmental conditions have yet to be reported.

The sequestration of chirality

An obvious and distinctive characteristic of homochiral protein and nucleic acid biopolymers is that they function generally within the enclosed environment of cells, which provide a membranous boundary structure which separates the intracellular components from the external environment. It has accordingly been postulated frequently that analogous but simpler enclosed environments must have been available and operative on the primitive Earth. In short, the array of extraterrestrial homochiral amino acids introduced to Earth as outlined above, for example, would not be expected to undergo further chemical evolution while floating freely in the oceans, but must have done so only after concentration and protection by sequestration into indigenous protocellular structures. We now examine briefly some of the suggestions which have been offered whereby this may have transpired.

Amphiphiles are molecules having a large hydrocarbon-like nonpolar moiety to which are attached one or more small polar groups. Lipid amphiphiles include such molecules as fatty acids and their triglyceride esters, sterols such as cholesterol and phospholipids such as lecithin. At an air-water interface amphiphiles are "surface active", that is they self-assemble by condensation into monolayers, with their hydrophilic polar groups extending into and their hydrophobic nonpolar groups out of the water surface. Above a certain critical concentration single-chain amphiphiles may aggregate further to form membranous micelles, in which hundreds of amphiphile units self-assemble further into a globular structure with their nonpolar hydrocarbon chains directed inward and their polar groups are directed outward toward the aqueus interface. At still higher concentrations micelles may fuse into multilamellar structures composed of bilayers [83]. Micelles have also been shown to be capable of self-replication. Here the increase in the number of micelle units is brought about by an appropriate chemical reaction occurring within the boundary of the micelle itself, which in turn leads to the production of more of the same amphiphile out of which the micelle was originally formed. Such self-replication has been demonstrated for both ordinary micelles in an aqueous medium [84] as well as for "reverse micelles" [85], which consist of minute droplets of water stabilized in an organic solvent by a layer of surfactant. Some of the prebiotic potentialities of such replicating membranous vesicles have been explored in further detail, and they have been characterized as "minimum protocells" [86].

A critical function of such lipid vesicle protocells in the prebiotic environment is their ability to sequester other molecules, including macromolecules, within the confines of the vesicle [86]. How might this occur? Deamer and Barchfeld addressed this problem experimentally in 1982 [87] by subjecting phospholipid vesicles (liposomes) to dehydration-rehydration cycles in the presence of either monomeric 6-carboxyfluorescein molecules or polymeric salmon sperm DNA molecules as extraneous solutes. The experiment was designed to mimic a prebiotic tidepool containing dilute dispersions of membranogenic phospholipids in the presence of external solutes, with the dehydration-rehydration cycles representing sequential dry and wet epochs in the primitive environment. The experiment showed that both the small molecule and the polymer were encapsulated in the vesicles formed after rehydration, with the extent of encapsulation increasing as the amount of original liposome was increased. The results were rationalized by the hypothesis that during the dehydration stage the original liposomes shrank and fused together, trapping the solutes between alternating lipid bilayers of an extended multilamellar sandwich. On rehydration the lamellae swelled to reform larger vesicles which had encapsulated a significant fraction of the original monomeric or polymeric solute. The authors also suggested that if both a template and a monomer were sequestered together, the close contact in the enclosed microenvironment might enhance the efficiency of template-directed replication, anticipating the later suggestion that membranous amphiphile vesicles might provide "an appropriate microenvironment for incorporation and evolution of polymer synthesis systems" [86]. We might speculate further that the dehydration-rehydration mechanism permitting the encapsulation of macromolecules within liposomes or similar vesicles might operate simultaneously with an e.e. amplification mechanism involving a partial-polymerization step (during dehydration) and a partial-hydrolysis step (during rehydration), to permit the evolution from amino acids of low e.e.s into homochiral polypeptides sequestered in the protected interiors of the vesicles. Such a possibility has not yet been investigated experimentally.

We thus see that the formation of membranous vesicle protocells, along with their ability to self-replicate and to encapsulate small- to polymeric-sized molecules, have been demonstrated by convincing experiments and thus could constitute realistic and plausible prebiotic processes on the primordial Earth. The crucial remaining question as to the potential availability of the requisite amphiphilic precursors in the primitive environment is one which has also been addressed experimentally by Deamer [87, 88]. In view of the extensive array of organic molecules which have been reported in the Murchison meteorite and other carbonaceous condrites [65, 66], it was logical to look into the uncontaminated. Murchison chondrite for the presence of such amphiphilic constituents. Samples of the meteorite were accordingly extracted with chloroform-methanol, and the extracts were fractionated by thin-layer chromatography. It was found that the components of some of the fractions displayed surface activity, forming monomolecular films at air-water interfaces, and were also able to self-assemble into membranous vesicles which could encapsulate polar solutes. These observations clearly demonstrated that amphiphiles plausibly available on the primitive Earth by meteoritic infall have the ability to self-assemble into the membranous vesicles of "minimum protocells". It is noteworthy that the existence of small L > D e.e.s in an array of homochiral amino acids as well as of amphiphiles with the ability to self-assemble and encapsulate solutes have both been found in the same extraterrestrial source – the Murchison meteorite.

A different type of system forming stable macroscopic membranes was described in 1993 by Zhang and co-workers in a study of oligopeptides having alternating hydrophobic and hydrophilic monomer residues [89]. They found that the addition of salt to the 16-unit polypeptide, (Ala-Glu-Ala-Glu-Ala-Lys-Ala-Lys)$_2$, which adopts a β-sheet secondary structure, caused it spontaneously to self-assemble to form stable macroscopic membranes, whereupon they suggested that such oligopeptide membranes could yield vesicles with an "enclosed environment for a primitive metabolism". It should be noted, however, that this potential encapsulation system not only lacks the broad generality of the previously discussed vesicle systems arising from amphiphiles, but the contrived and stringent experimental conditions necessary for its realization clearly disqualify it is being a realistic mechanism in the prebiotic environment.

Conclusions

Our objective in this review has been to survey critically the hypotheses which have been advanced in the past to explain the origin and ubiquity of the homochiral molecules which distinguish contemporary biopolymers: L-amino acids in proteins and D-sugars in the nucleic acids, RNA and DNA. In doing so we have evaluated each proposed theory or mechanism as to both its experimental validity as well as its potential viability in the known [16–19] violent and inhospitable environment characteristic of the primordial Earth.

After emphasizing the crucial role of homochirality and enantiomeric homogeneity in evolving biomolecules, terrestrial hypotheses for the origin of chirality have been surveyed in two classes, (i) *random* mechanisms, which have an equal probability of producing either enantiomer, and (ii) *determinate* mechanisms, where the resulting chirality is predetermined. Random terrestrial mechanisms which involve spontaneous symmetry breaking of the racemic prebiotic organic medium by autocatalytic processes [9, 22–29], as well as mechanisms involving solid-state lattice-controlled asymmetric syntheses [30], even when validated experimentally, have been found to require such contrived and manipulated experimental conditions as to invalidate their potential for prebiotic relevance. Alleged asymmetric phenomena involving clay minerals have been found to be factitious, and reported stereoselective adsorption and polymerization processes relying on (+)- or (–)-quartz, which qualify as stochastic because of the random terrestrial distribution of enantiomeric quartz crystals, have been shown either to be erroneous or to be vitiated in the prebiotic environment because of the anhydrous media required to achieve valid stereoselective effects. As to determinate terrestrial mechanisms, experiments claiming the efficacy of electric, magnetic and gravitational fields for generating a particular chirality have been found to be either questionable theoretically or based on subsequently disclosed experimental errors. Determinate mechanisms based on parity violation at the elementary particle level include those based on β-decay, parity-violating energy differences, and Bose condensation phase changes [38]. Experiments involving β-decay electrons or their Bremsstrahlen, as well as other spin-polarized chiral elementary particles, have led to either negative or nonreproducible results, while PVED mechanisms and Bose condensations remain only controversial theoretical assertions, with efforts to substantiate them experimentally so far leading to consistently negative results [36, 37, 39]. The only determinate mechanisms which have proved capable of producing valid and reproducible stereoselective effects have been asymmetric syntheses, asymmetric photoequilibrations and asymmetric photolyses induced by circularly polarized light [33]. Of these, only partial asymmetric photolysis has any intrinsic potential for prebiotic applicability, but even this efficient mechanism, circumscribed as we have seen, could hardly be effectual on the primitive Earth. We thus conclude and emphasize that even though a positive experimental result may be observed in the laboratory for a potentially enantioselective process, this by no means automatically endows that process with relevance for the prebiotic terrestrial origin of chirality.

In view of the inadequacy of the random and determinate mechanisms discussed above to account for the terrestrial origin of chiral molecules having discernible e.e.s, the comprehensive hypothesis illustrated in Figure 2, an extension of the original suggestions of Rubenstein and coworkers [54], was proposed for the extraterrestrial origin of chiral molecules and their transport to Earth. While the sequence of individual steps in Figure 2 is comprehensive, ranging from the initial supernova event to the final sequestration of chirally pure polymers in terrestrial protocells, each step in the sequence is supported by theoretical, observational or experimental evidence. Here we shall review briefly the documentation for each of the 10 steps in Figure 2.

Step (1), the explosion of a supernova and the formation of neutron stars is discussed by Bethe and Brown [49], and step (2), the synchrotron radiation emanating from a neutron star by Oort [50]. Step (3), concerning the spectrum and the presence of RCPL and LCPL in the components of synchrotron radiation from laboratory synchrotron sources is documented by Winick [51]. Step (4), the formation of racemic mantle constituents in interstellar grain mantles is supported by the experiments of Briggs and coworkers [53], where complex organic mixtures were produced on simulated grain mantels by ultraviolet irradiation of frozen small molecule "ices". The asymmetric photolysis of such racemic molecules in grain mantles by ultraviolet CPL at interstellar temperatures, step (5), is authenticated by the model experiments of Greenberg [55]. Step (6), the cold accretion of interstellar grains by Earth as the solar system traverses interstellar molecular clouds, is discussed by Greenberg and others [52], while the coalescence of interstellar grains into comets and asteroids, step (7), is documented by Greenberg and co-workers [43, 53]. Nondestructive impacts with Earth, step (8), are documented by Chyba and Greenberg and their co-workers [57–59], and the ultimate introduction of homochiral amino acids with small L > D e.e.s by Engel and Cronin and co-workers [67–70]. Model experiments documenting the possibility of step (9), the abiotic amplification of amino acids during polymerization, have been performed by Bonner and co-worker [78, 80] and by Brach and Spach [79]. Finally, step (10), the sequestration of small and polymeric molecules by encapsulation into amphiphile vesicles, has been verified by the experiments of Deamer and co-workers [86–88].

We thus see that each of the 10 sequential steps in the extended hypothesis of Figure 2 has been substantiated by either theoretical, observational or experimental documentation. We might therefore rationalize the existence of L-amino acids and the lack of D-amino acids in our biosphere by the assumption that the CPL available for the asymmetric photolysis of the racemic constituents of interstellar grain mantles, step (5), in the part of the galaxy where our solar system was formed was such as to produce L-amino acids in the mantles, which ultimately resulted in L-homochiral proteins on the primitive Earth. But other aspects of the overall picture regarding our homochiral terrestrial biomolecules remain obscure and tainted with countless unanswered questions. Although the scenario of Figure 2 may be satisfactory for proteins derived from L-amino acids, there is nothing in the picture to rationalize the exclusive presence of D-sugars in the nucleic acids of our biosphere. Until more is known about the actual prebiotic evaluation and chirality of the precursors leading to ribose and deoxyribose, we are hardly in a position to speculate on the origin of the D homochirality of these sugars.

References

1. Mislow, K. and Bickart, P. (1976/77) An epistemological note on chirality. *Israel J Chem* 15:1–6.
2. MacDermott, A.J. (1966) The weak force and SETH: the search for extra-terrestrial homochirality. In: D.B. Cline (ed): *Physical origin of homochirality in life*. American Institute of Physics, Woodbury, New York, 241–254.
3. Halpern, B., Westley, J.W., Levinthal, E.C. and Lederberg, J. (1966) The Pasteur probe: an assay for molecular asymmetry. *Life Sci Space Res* 5:239–249.
4. Bonner, W.A. (1996) The quest for chirality. In: D.B. Cline (ed): *Physical origin of homochirality in life*. American Institute of Physics, Woodbury, New York, 17–49.
5. Bonner, W.A. (1991) The origin and amplification of biomolecular chirality. *Orig Life Evol Biosphere* 21: 59–111.
6. Bonner, W.A. (1988) Origins of chiral homogeneity in nature. In: E.L. Eliel, S.H. Wilen (eds): *Topics in stereochemistry*. Wiley, New York, 1–96.
7. Bonner, W.A. (1972) Origins of molecular chirality. In: C. Ponnamperuma (ed): *Exobiology*. North-Holland, Amsterdam, 170–234.
8. Miller, S.L. and Orgel, L.E. (eds) (1974) *The origin of life on Earth*. Prentice-Hall, Englewood Cliffs, New Jersey, 166–174.

9. Avetisov, V. and Goldanskii, V. (1997) Mirror symmetry breaking at the molecular level. *Proc Natl Acad Sci USA* 93: 11435–11442.
10. Joyce, G.F., Visser, G.M., van Boeckel, C.A.A., van Bloom, J.H., Orgel, L.E. and van Westrenen, J. (1984) Chiral selection in poly(C)-directed synthesis of oligo(G). *Nature* 310: 602–604.
11. Goldanskii, V.I., Anikin, S.A., Avetisov, V.A. and Kuzmin, V.V. (1987) On the decisive role of chiral purity in the origination of the biosphere and in its possible destruction. *Comm Mol Cell Biophys* 4: 79–98.
12. Bada, J.L. and Miller, S.L. (1987) Racemization and the origin of optically active organic compounds in living organisms. *Bio Systems* 20: 21–26.
13. Bonner, W.A., Blair, N.E. and Lemmon, R.M. (1980) The radioracemization of amino acids by ionizing radiation: geochemical and cosmochemical implications. In: P.E. Hare, T.C. Hoering, K. King, Jr. (eds): *Biogeochemistry of amino acids*. Wiley, New York, 357–374.
14. Keszthelyi, L., Czégé, J., Fajszi, C., Pósfai, J. and Goldanskii, V.I. (1979) Racemization and the origin of asymmetry of biomolecules. In: D.C. Walker (ed.): *Origins of optical activity in nature*. Elsevier, New York, 229–244.
15. Shapiro, R. (1986) *Origins – a skeptic's guide to the creation of life on Earth*. Summit Books, New York, 98–116.
16. Waldrop, M.M. (1990) Good-bye to the warm little pond. *Science* 250: 1078–1080.
17. Chyba, C.F. (1992) The violent environment of the origin of life. In: J and K Tran Thanh Van, J.C. Mounolou, J. Schneider, C. McKay (eds): *Frontiers of life*. Editions Frontières, Cedex, France 97–104.
18. Maher, K.A. and Stevenson, D.J. (1988) Impact frustration of the origin of life. *Nature* 331: 612–614.
19. Oberbeck, V.R. and Fogleman, G. (1990) Impact constraints on the environment for chemical evolution and the continutiy of life. *Orig Life Evol Biosphere* 20: 181–195.
20. Sagan, C. (1974) The origin of life in a cosmic context. *Orig Life* 5: 497–505.
21. Overbeck, V.R. and Fogleman, G. (1989) Estimates of the maximum time required to originate life. *Orig Life Evol Biosphere* 19: 549–560.
22. Bolli, M., Micura, R. and Eschenmoser, A. (1997) Pyranosyl-RNA: chiroselective self-assembly of base sequences by ligative oligomerization of tetranucleotide-2′,3′-cyclophosphates (with a commentary concerning the origin of biomolecular homochirality). *Chem Biol* 4: 309–320.
23. Frank, F.C. (1953) On spontaneous asymmetric synthesis. *Biochim Biophys Acta* 11: 459–463.
24. Calvin, M. (1969) *Chemical evolution*. Oxford University Press, New York, 149–152.
25. Avetisov, V.A. (1996) Spontaneous mirror symmetry breaking via enantioselective autocatalysis. In: D.B. Cline (ed.): *Physical origin of homochirality in life*. American Institute of Physics, Woodbury, New York, 141–150.
26. Lippmann, D. and Pattanayak, A. (1996) Symmetry breaking by autocatalysis. In: D.B. Cline (ed.): *Physical origin of homochirality in life*. American Institute of Physics, Woodbury, New York, 151–155.
27. Buhse, T., Lavabre, D., Micheau, J. and Thiemann, W. (1993) Chiral symmetry breaking: experimental results and computer analysis of a liquid-phase autoxidation. *Chirality* 5: 341–345.
28. Shibata, T., Morioka, H., Hayase, T., Choji, K. and Soai, K. (1996) Highly enantioselective catalytic asymmetric automultiplication of chiral pyrimidyl alcohol. *J Am Chem Soc* 118: 471–472.
29. Bolm, C., Bienewald, F. and Seger, A. (1996) Asymmetric autocatalysis with amplification of chirality. *Angew Chem Int Ed Engl* 35: 1657–1659.
30. Vaida, M., Popovitz-Biro, R., Leiserowith, L. and Lahav, M. (1991) Probing reaction pathways via asymmetric transformations in chiral and centrosymmetric crystals. In: V. Ramamurthy (ed.): *Photochemisty in organized and constrained media*. VCH Publishers, New York, 248–302.
31. Barron, L.D. (1994) Can a magnetic field induce absolute asymmetric syntheses? *Science* 266: 1491–1492.
32. Barron, L.D. (1996) True and false chirality, CP violation and the breakdown of microscopic reversibility in chiral molecular and elementary particle processes. In: D.B. Cline (ed.): *Physical origin of homochirality in life*. American Institute of Physics, Woodbury, New York, 162–182.

33. Bonner, W.A. and Rubenstein, E. (1990) Photochemical origins of biomolecular chirality. In: C. Ponnamperuma, F.R. Eirich (eds) *Prebiological self-organization of matter*. A. Deepak, Hampton, Virginia, 35–50.
34. Yamagata, Y. (1966) A hypotheseis for the asymmetric appearance of biomolecules on Earth. *J Theor Biol* 11:495–498.
35. Kondepudi, D.K. (1994) Selection of handedness in prebiotic chemical processes. In: *Discovery of weak neutral currents*. AIP Conf. Proc. 300:491–498.
36. Bonner, W. (1996) Enantioselective autocatalysis. IV. Implications for parity violation effects. *Orig Life Evol Biosphere* 26:27–46.
37. Kondepudi, D.K., Bullock, K.L., Digits, J.A. and Yarborough, P.D. (1995) Stirring rate as a critical paramater in chiral symmetry breaking crystallization. *J Am Chem Soc* 117: 401–404.
38. Salam, A. (1991) The role of chirality in the origin of life. *J Mol Evol* 33:105–113.
39. Figureau, A., Duval, E. and Boukenter, A. (1995) Can biological homochirality result from a phase transition? *Orig Life Evol Biosphere* 25:211–217.
40. Oró, J. (1961) Comets and the formation of biochemical compounds on the primitive Earth. *Nature* 190:389–390.
41. Delsemme, A.H. (1992) Cometary origin of carbon, nitrogen and water on the Earth. *Orig Life Evol Biosphere* 21:279–298.
42. Huebner, W.F. and Boice, D.C. (1992) Comets as possible sources of prebiotic molecules. *Orig Life Evol Biosphere* 21:299–315.
43. Greenberg, J.M. and Menoza-Gomez, C.X. (1992) The seeding of life by comets. *Adv Space Res* 12:169–180.
44. Marcus, J.N. and Olsen, M.A. (1991) Biological implications of organic compounds in comets. In: R.L. Newburn, M. Neugebauer and J. Rahe (eds): *Comets in the post-Halley era*. Kluwer, Dordrecht, 439–462.
45. Love, S.G. and Brownlee, D.E. (1993) A direct measurement of the terrestrial mass accretion rate of cosmic dust. *Science* 262:550–553.
46. Whittet, D.C.B. (1997) Is extraterrestrial organic matter relevant to the origin of life on Earth? *Orig Life Evol Biosphere* 27:249–262.
47. Cyba, C. and Sagan, C. (1992) Endogenous production, exogenous delivery and impact-shock synthesis of organic molecules: an inventory for the origins of life. *Nature* 355: 125–132.
48. Rubenstein, E., Bonner, W.A., Noyes, H.P. and Brown, G.S. (1983) Supernovae and life. *Nature* 306:118.
49. Bethe, H.A. and Brown, G. (1985) How a supernova explodes. *Sci Am* 252:60–68.
50. Oort, J.H. (1957) The Crab nebula. *Sci Am* March: 53–69.
51. Winick, H. (1980) Properties of synchrotron radiation. In: H. Winick and S. Doniach (eds): *Synchrotron radiation research*. Plenum, New York, 11–25.
52. Greenberg, J.M. and Mendoza-Gómez, C.X. (1993) Interstellar dust evolution: a resorvoir of prebiotic molecules. In: J.M. Greenberg, C.X. Mendoza-Gómez and V. Pirronello (eds): *The chemistry of life's origins*. Kluwer, Dordrecht, 1–32.
53. Briggs, R., Ertem, G., Ferris J.P., Greenberg, J.M., McCain, P.J., Mendoza-Gómez, C.X. and Schutte, W. (1992) Comet Halley as an aggregate of interstellar dust and further evidence for the photochemical formation of organics in the interstellar medium. *Orig Life Evol Biosphere* 22:287–307.
54. Bonner, W.A. and Rubenstein, E. (1987) Supernovae, neutron stars and biomolecular chirality. *Bio Systems* 20:99–111.
55. Greenberg, J.M. (1996) Chirality in interstellar dust and in comets: life from dead stars. In: DB Cline (ed.): *Physical origin of homochirality in life*. American Institute of Physics, Woodbury, New York, 185–210.
56. Greenberg, J.M. and Li, A. (1996) From cosmic formation of chiral bioorganics in interstellar dust to comet seeds of life's origin. Are 10^{25} chances enough? In: J. Chela-Flores and F. Raulin (eds): *Chemical evolution: physics of the origin and evolution of life*. Kluwer, Dordrecht, 51–71
57. Chyba, C.F. (1990) Impact delivery and erosion of planetary oceans in the early inner solar system. *Nature* 343:129–133.
58. Cyby, C.F., Thomas, P.J., Brookshaw, L. and Sagan, C. (1990) Cometary delivery of organic molecules to the early Earth. *Science* 249:366–373.

59. Greenberg, J.M., Kouchi, A., Niessen, W., Irth, H., van Paradijs, J., de Groot, M. and Hermsen, W. (1994) Interstellar dust, chirality, comets and the origins of life: life from dead stars? *J Biol Phys* 20:61–70.
60. Clark, B.C. (1988) Primeval procreative comet pond. *Orig Life Evol Biosphere* 18:209–238.
61. Kerr, R.A. (1997) Spots confirmed, tiny comets spurned. *Science* 276:1333–1334.
62. Frank L.A., Sigwarth, J.B. and Craven, J.D. (1986) On the influx of small comets ino the Earth's upper atmosphere II. Interpretation. *Geophys Res Lett* 13:307–310.
63. Steigerwald, W.A. (1997) Polar spacecraft images support theory of interplanetary snowballs spraying Earth's upper atmosphere. *NASA News*, 28 May. Goddard Space Flight Center, Office of Public Affairs, Greenbelt, MD.
64. Kerr, R.A. (1988) In search of elusive little comets. *Science* 240:1403–1404.
65. Cronin, J.R. and Chang, S. (1993) Organic matter in meteorites: molecular and isotopic analyses of the Murchison meteorite. In: J.M. Greenberg, C.X. Mendoza-Gómez and V. Pirronello (eds): *The chemistry of life's origins*. Kluwer, Dordrecht, 209–258.
66. Kvenvolden, K., Lawless, J., Pering, K., Peterson, E., Flores, J., Ponnamperuma, C., Kaplan I.R. and Moore, C. (1970) Evidence for extraterrestrial amino-acids and hydrocarbons in the Murchison meteorite. *Nature* 228:623–626.
67. Engel, M.H. and Nagy, B. (1982) Distribution of enantiomeric composition of amino acids in the Murchison meteorite. *Nature* 296:837–840.
68. Engel, M.H., Macko, S.A. and Silfer, J.A. (1990) Carbon isotope composition of individual amino acids in the Murchison meteorite. *Nature* 348:47–49.
69. Epstein, S., Krishnamurthy, R.V., Cronin, J.R., Pizzarello, S. and Yuen, G.U. (1987) Unusual stable isotope ratios in amino acid and carboxylic acid extracts from the Murchison meteorite. *Nature* 326:477–479.
70. Cronin, J.R. and Pizzarello, S. (1997) Enantiomeric excesses in meteoritic amino acids. *Science* 275:951–954.
71. Bonner, W.A., Blair, N.E., Lemmon, R.M., Flores, J.J. and Pollock, G.E. (1979) The radioracemization of isovaline: cosmochemical implications. *Geochim cosmochim acta* 43: 1841–1846
72. Kagan, H.P. and Fiaud, J.C. (1988) Kinetic resolutions. In: E.L. Eliel and S.H. Wilen (eds): *Topics in stereochemistry*, vol. 18. Wiley, New York, 249–330.
73. Addadi, L., Berkovitch-Yellin, Z., Weissbuch, I., van Mil, J., Shimon, L.J.W, Lavas, M. and Leiserowitz, L. (1985) Growth and dissolution of organic crystals with "tailor-made" inhibitors – implications in stereochemistry and materials science. *Angew Chem Int Ed Engl* 24:466–485.
74. Thiemann, W. and Teutsch, H. (1990) Possible amplification of enantiomeric excesses through structural properties of liquid crystals – a model for the origin of optical activity in the biosphere. *Orig Life Evol Biosphere* 20:121–126.
75. Morowitz, H.J. (1969) A mechanism for the amplification of fluctuations in racemic mixtures. *J Theor Biol* 25:491–494.
76. Thiemann, W. (1974) Disproportionation of enantiomers by precipitations. *J Mol Evol* 4:85–97.
77. Wald, G. (1957) The origin of optical activity. *Ann NY Acad Sci* 69:352–368.
78. Blair, N.E. and Bonner, W.A. (1980)Experiments on the amplification of optical activity. *Orig Life* 10:225–263.
79. Brack, A. and Spach, G. (1979) β-Structures of polypeptides with L- and D-residues. Part I. Synthesis and conformational studies. *J Mol Evol* 13:35–46.
80. Blair, N.E., Dirbas, F.M. and Bonner, W.A. (1981) Stereoselective hydrolysis of leucine oligomers. *Tetrahedron* 37:27–29.
81. Blair, N.E. and Bonner, W.A. (1981) A model for the enantiomeric enrichment of polypeptides on the primitive Earth. *Orig Life* 11:331–335.
82. Brack, A. and Spach, G. (1980) β-Structures of polypeptides with L- and D-residues. Part III. Experimental evidences for enrichment in enantiomer. *J Mol Evol* 15:231–238.
83. Deamer, D.W. (1986) Role of amphiphilic compounds in the evolution of membrane structure on the early Earth. *Orig Life* 17:3–25.
84. Bachman, P.A., Luisi, P.L. and Lang, J. (1992) Autocatalytic self-replicating micelles as models for prebiotic structures. *Nature* 257:57–59.
85. Bachmann, P.A., Luisi, P.L. and Lang, J. (1991) Self-replicating reverse micelles. *Chimia* 45:266–268.

86. Morowitz, H.J., Heinz, B. and Deamer, D.W. (1988) The chemical logic of a minimum protocell. *Orig Life Evol Biosphere* 18:281–287.
87. Deamer, D.W. and Barchfeld, G.L. (1982) Encapsulation of macromolecules by lipid vesicles under simulated prebiotic conditions. *J Mol Evol* 18:203–206.
88. Deamer, D.W. and Pashley, R.M. (1989) Amphiphilic components of the Murchison carbonaceous chondrite: surface properties and membrane formation. *Orig Life Evol Biosphere* 19:21–38.
89. Zhang, S., Holmes, T., Lockshin, C. and Rich, A. (1993) Spontaneous assembly of a self-complementary oligopeptide to form a stable macroscopic membrane. *Proc Natl Acad Sci USA* 90:3334–3338.

Subject index

Detailed information about all our titles also available on the internet:

http://www.birkhauser.ch

Check our Highlights for new and notable titles selected monthly in each field

Oncogenesis
Molecular Biology
Cell Biology

MCBU
Molecular and Cell Biology Updates

Bannasch P., Deutsches Krebsforschungszentrum, Heidelberg,Germany / **Kanduc D.,** University of Bari / **Papa S.,** University of Bari / **Tager J.M.,** University of Amsterdam, The Netherlands (Ed.)

Cell Growth and Oncogenesis

1998. 314 pages. Hardcover
ISBN 3-7643-5727-4

Rapid progress has been made in our understanding of the molecular mechanisms of cell growth and oncogenesis during the past decade. This book comprises recent results on the regulation of cell growth in normal and neoplastic tissues by growth factors including hormones, and by the activation and inactivation of oncogenes and tumor suppressor genes, respectively. Special attention has been given to the presentation of the frequently neglected close correlation between changes in signal transduction and metabolism pathways during oncogenesis.

Table of contents:

BioSciences with Birkhäuser

(Prices are subject to change without notice. 8/98)

For orders originating from all over the world except USA and Canada:

Birkhäuser Verlag AG
P.O. Box 133
CH-4010 Basel / Switzerland
Fax: +41 / 61 / 205 07 92
e-mail: orders@birkhauser.ch

For orders originating in the USA and Canada:

Birkhäuser Boston, Inc.
333 Meadowland Parkway
USA-Secaurus, NJ 07094-2491
Fax: +1 / 201 348 4033
e-mail: orders@birkhauser.com